HISTORY 4° CELSIUS

HIST

Specters of the Atlantic
Specters of the Atlantic: Finance Capital, Slavery, and the Philosophy of History, volume 1
History 4° Celsius: Search for a Method in the Age of the Anthropocene, volume 2

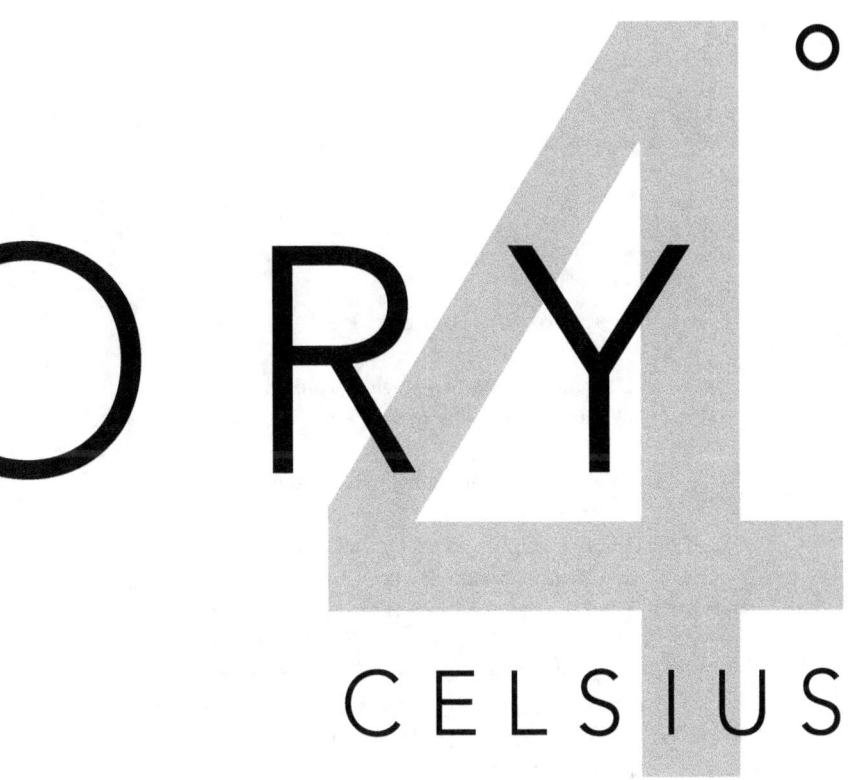

HISTORY 4° CELSIUS

Search for a Method in the Age of the Anthropocene

IAN BAUCOM

Duke University Press   Durham and London   2020

© 2020 Duke University Press
All rights reserved
Designed by Courtney Leigh Richardson
Typeset in Whitman and Avenir by Copperline Book Services

Library of Congress Cataloging-in-Publication Data
Names: Baucom, Ian, [date] author.
Title: History 4° celsius : search for a method in the age
of the Anthropocene / Ian Baucom.
Other titles: History 4 degrees celsius | Theory in forms. Description:
Durham : Duke University Press, 2020. | Series: Theory in forms |
Includes bibliographical references and index. Identifiers: LCCN
2020005103 (print) | LCCN 2020005104 (ebook) ISBN
9781478007876 (hardcover)
ISBN 9781478008392 (paperback)
ISBN 9781478012030 (ebook)
Subjects: LCSH: Slave trade—Ghana—History. | Climatic changes—
Economic aspects—History. | Climatic changes—
Social aspects—History. | Capitalism—Environmental aspects—
History. | Capitalism—Social aspects—History. | Geology,
Stratigraphic—Anthropocene.
Classification: LCC HT1394.G48 B38 2020 (print) |
LCC HT1394.G48 (ebook) | DDC 306.3/6209667—dc23
LC record available at https://lccn.loc.gov/2020005103
LC ebook record available at https://lccn.loc.gov/2020005104

COVER ART: *Seascape*. Courtesy of Getty Images/
Claudio Pennini/EyeEm

This book is dedicated to Srinivas Aravamudan: incomparable scholar, world dean and diplomat of the humanities, beloved friend, older brother. It would not exist without him. Nor would I.

CONTENTS

Acknowledgments · ix

1. **Of Forces and Forcings** · 1
2. **History 4° Celsius: Search for a Method** · 35
3. **The View from the Shore** · 73
Coda. **The Youngest Day** · 110

Notes · 119
Bibliography · 131
Index · 137

ACKNOWLEDGMENTS

I have more people to thank for their thought, insight, encouragement, and friendship as I've worked on this book than I can adequately acknowledge. Two immediately stand out for having prompted me into thought: Debjani Ganguly, whose astonishing conference on the Anthropocene Humanities, friendship, and intellectual companionship over so many years have so deeply shaped and inspired me; and Dipesh Chakrabarty, without whose writing and conversation I could not have begun to imagine this work. Along the way from beginning to consider this project to completing the text, I've been given the gift of touchstone friendships and critical co-travelers. Srinivas, Jim Chandler, Achille Mbembe, Ranji Khanna, Charlie Piot, Sarah Nuttall, Anne Allison, Ken Wissoker, Willis Jenkins, Karen McGlathery, Amitav Ghosh, Tim Murray, Renate Ferro, Amanda Anderson, Laurent Dubois, Premesh Lalu, Kathy Woodward, Gary Tomlinson, Rosi Braidotti, Sara Guyer, Julia Adeney Thomas, Ato Quayson, Matthew Omelsky, Ainehi Edoro, Chad Wellmon, Matthew Burtner, Chris Chia, Raffaele Laudani, Steve Arata, Andy Stauffer, Bruce Holzinger, Jim Seitz, John Parker, all the students in my Anthropocene seminar at the School for Criticism and Theory: I am more profoundly in your debt than you possibly know. For those I've failed to name, please know that I know you, and everything you've made possible. Last in name but first in my love: Wendy, Gabriel, Leah, Kiran, Camden, Tristan—not one thought, hope, moment of joy, or dream for the future without you.

1. **Of Forces and Forcings**

The strip of land between Keta and Axim on the coast of Ghana has come to occupy a singular place in recent histories of the modern world. There is a bleak reason for that. Or, more precisely, a forty-eight-count bundle of reasons. As Stephanie Smallwood, Saidiya Hartman, and Bayo Holsey, among others, have reminded us, the Ghanaian coast is more than a coast.[1] It is an archipelago of no fewer than forty-eight gateposts onto a living diasporic modernity, forty-eight "factories" of the modern scattered along this scant 335 miles of shoreline. Built over the course of the seventeenth and eighteenth centuries by successive waves of Portuguese, Spanish, Dutch, English, Danish, and Swedish merchants, the forts were castles turned trading posts and slaveholding barracoons: "factories" in which hundreds of thousands of Africans captured for sale to the Americas saw their lives rendered subject to "practices" of restraint, measurement, and violence "designed to promote the pretense that human beings could convincingly play the part of their antithesis—bodies animated only by others' calculated investment in their physical capacities."[2] As file after file of captives was marched to this factory-crowded Atlantic littoral—year after year, decade after decade, century after century—the culture and machinery of transatlantic slavery continuously transformed Ghana's shore into what Smallwood appositely calls a "stage" for the constitutive "activities and practices" of the modern. That stage was peopled by the children, men, and women who were drawn from across the interior of the continent, held in the dungeons, suffering the violence of their fraudulent commodification and, even in that suffering, beginning to craft the practices of survival, the repertoires of creolization, and the orientations toward a long and still unfinished politics of freedom, which, in Paul Gilroy's terms, made of them, as they made of themselves in the most constrained of situations, a vanguard of the world's first "truly modern people."³

Eddying around these captives and these factories were other spiraling circles of actors for whom the Ghanaian "Gold Coast" had become an equally powerful vortex of the modern: the Asante, Akwamu, and other political elites organizing their war-making projects, their strategies for plundering prisoners from the hinterlands, and their unfolding state forms in relation to the relentless demand for saleable labor in the Caribbean and in North and South America, much as their political counterparts and distant co-beneficiaries in Spain, the Dutch Republics, France, and England were crafting their post-Westphalian war-making projects and their own unfolding state forms in relation to the social and market worlds of the West African littoral and the transatlantic colonies.⁴ Allied to, servicing, and profiting from these political actors were the merchant agents of the seventeenth- through nineteenth-century Dutch and English chartered companies, and other European slave-trading nationals. For these commercial agents, the factories were not merely storehouses of captive labor but, in Giovanni Arrighi's terms, "spaces-of-flow" for modern capital's ever more globe-crossing routes of circulation, "Bizensone Fairs" (as Saidiya Hartman has also suggested) of an Atlantic cycle of capital accumulation drawing together European textile manufacturing, the plantation economies of the "New World," the world trade in gold bullion, and the speculative revolutions in modern finance capital that could not have come into being without the slave trade and the array of financial instruments, bills of exchange, joint-stock schemes, credit networks, and insurance mechanisms it helped launch.⁵

Of the forty-eight slave factories strung along the Ghanaian coast, one has previously been of particular interest to me. In 1781 Captain Luke Collingwood called at Fort William in Anomabo, where he loaded the majority of the slaves aboard his ship, the *Zong*, prior to setting sail for Jamaica. Before he made port in the Caribbean, he determined to drown rather than sell 132 of those stolen captives and to claim the thirty pounds per head that the ship's marine insurance contract had established as their valuation. If, as I have argued in the first volume of this series, that decision—that violent conversion of human life into a legally enforceable quantum of speculative value—proved emblematic of the coming into being of a hypercapitalized long contemporaneity stretching from the late eighteenth century into our present, then it is vital to recall that what transpired aboard the *Zong* could not have occurred without what had first taken place at Fort William.[6] The *Zong*'s was but one voyage, embarking from but one of the dozens of Gold Coast slave factories over a multicentury period of time, and the history of its Atlantic passage—as the history of any slave voyage—is linked to and emerged from a vast factory system taking shape on West Africa's Atlantic shore, a system encompassing not only Fort William but Fort Santo Antonio, Fort Metal Cross, Fort San Sebastian, Fort Batenstein, Fort Coenraadsburg, Fort Amsterdam, Fort Lijdzaamheid, Fort Goede Hoop, Fort Orange, Ussher Fort, Teshie Fort, Fort Gross-Friedrichsburg, Fort Prinzenstein, Fort Apollonia, Osu Castle, Cape Coast Castle, Elmina Castle, and the scores of other factories stretching all along the coast of West Africa, siphoning not hundreds or thousands but millions of lives.

The camp, Giorgio Agamben has argued, must be regarded as an emblem of the modern, as embodying a definitive modern *nomos* of the earth through the violent and defining relationship it establishes between sovereign power and bare, eradicable life.[7] To the camp we must add another emblematic institution of the modern, another site or archipelago of sites, drawing together sovereign power, disposable life, transnational capital, the culture of créolité, and the politics of survival: the slave factories of the Atlantic coast of Africa. Individually, each of these places, every one—Fort William, Fort Amsterdam, Elmina Castle—resides in the history of the world as a place of singular violence. Collectively, in their arrayed series, sustaining and replicating themselves over centuries of time, they reappear as representative places: places in which the key nomological codes of cultural, social, political, and capital life governing one of the dominant stages of modernity have been written; places of stunning constraint, which, for that very reason, also became staging grounds for one of the world's great, enduring, and still incomplete politics of freedom; emblematic places of the contending forces and force fields of the modern.

That is one understanding we can and must have of the coast of Ghana: an understanding of its fundamental centrality to the history of the modern world.

There is now another conception of that coast with which we must also come to terms, a conception less of the shore's remaking of and by the forces of modern history than of its reconstitution by the forcings of climate change. That new understanding is thrown into sharp relief by a single image— one of a series of ten photographs collectively entitled "We Were Once Three Miles from the Sea," taken by the Ghanaian photographer Nyani Quarmyne

Figure 1.1. Collins Kusietey, "We Were Once Three Miles from the Sea." © Nyani Quarmyne/Panos Pictures.

of the villages of Totope, Azizakpe, Akplabanya, Lolonyakope, and Azizanya near the mouth of the Volta River, along the coast from Fort William, between March 2010 and February 2011. Seven-year-old Collins Kusietey stands in a roofless, plaster-walled house, half-filled with beach sand that the encroaching waves and advancing shoreline have piled into his home.

He is staring directly at us, half dressed, framed by a gold-painted window aperture. The image is profoundly unsettling, both beautiful and overwhelmed by a sense of wrongness pervading its carefully ordered visual space: the wrongness of the presence of that sand in his house; the wrongness of his extreme precariousness and vulnerability in life and before the lens of the camera; the wrongness of the ocean, somewhere back behind him, slowly rising, worldwide, millimeter by millimeter, as the glaciers and ice sheets melt, but here in Totope, near Ada, on the shore of one of the slave forts of the transatlantic slave trade, leaping suddenly, massively, devastatingly higher in one of those terrible local asymmetries of a global condition.

As we regard this image, what do we see?

The Black Atlantic? Or the Atlantic?

The forces of history? Or the forcings of what we have recently come to understand as the Anthropocene?

While I was gathering notes for this book, the *New York Times* published a front-page story under the bleak headline "Heat-Trapping Gas Passes Milestone, Raising Fears." In the dire news it communicates, the efficiency with which it shares that news, and the hybrid mathematics of time it draws on for that communiqué, the story provides an unfortunately perfect précis of many of my core concerns. "The level of the most important heat-trapping gas in the atmosphere, carbon dioxide," the *Times* account opens, "has passed a long-feared milestone, scientists reported on Friday, reaching a concentration not seen on the earth for millions of years. Scientific monitors reported that the gas had reached an average daily level that surpassed 400 parts per million."[8] Following those opening sentences, the news gets worse, and worse in a distinctive way, not only tending toward a catastrophic result, but moving in that ruinous future direction through a distinctive marshaling of moments, periods, and timescales that have made climate reporting one of the outer frontiers of a new theory of historical time.

The story continues:

> The best available evidence suggests the amount of the gas in the air has not been this high for at least three million years. Carbon dioxide above 400 parts per million was first seen in the Arctic last year. . . . [but] the average reading for an entire day surpassed that level . . . for the first time in the 24 hours that ended at 8 p.m. Eastern Daylight Time [on May 9, 2013]. . . . From studying air bubbles trapped in Antarctic ice, scientists know that going back 800,000 years, the carbon dioxide level oscillated in a tight band, from about 180 parts per million in the depths of ice ages, to about 280 during the warm periods between. . . . For the entire period of human civilization, roughly 8,000 years, the carbon dioxide level was relatively stable near that upper bound. But the burning of fossil fuels has caused a 41 percent increase in the heat-trapping gas since the Industrial Revolution. Indirect measurements suggest that the last time the carbon dioxide level

was this high was at least three million years ago, during an epoch called the Pliocene. Geological research shows that the climate then was far warmer than today, the world's ice caps were smaller, and the sea level might have been as much as 60 or 80 feet higher. Experts fear that humanity may be precipitating a return to such conditions—except this time, billions of people are in harm's way.[9]

These are not the sorts of dilemmas that as a literary scholar I was trained to address. As I completed my graduate education and began my career in the mid-1990s, the looming planetary crisis of climate change had not yet become a matter of broad common recognition and concern. Even within the deeply historically minded field of postcolonial studies, the modes of conceiving of historical time that this story treats as virtual commonplaces were largely inconceivable—both in their dizzying jumps between temporal scales (from a particular hour on a particular day; to the approximately sixty years in which we have been keeping accurate measurements of carbon dioxide emissions; to the segment of time since the Industrial Revolution; to "the entire [8,000-year] period of human civilization"; to the 800,000-year history of Antarctic Ice; to the three million years since the epoch of the Pliocene) and in the theory of historical periodization enabling those scale-shifting moves. While I have for some time accepted the force of Frederic Jameson's dictum that "we cannot not periodize," until recently it would not have occurred to me that postcolonial study, critical theory, or the humanities disciplines in general needed to periodize in relation to capital *and* also to carbon, in postmodernities *and* in parts per million, in dates *and* in degrees Celsius.[10]

As the crisis of climate change has become as starkly apparent as these news accounts reveal, I have joined other scholars in the humanities wrestling with precisely such questions. Like many colleagues in postcolonial studies, I have been doing so in relation to the pioneering work of Rob Nixon, Ramachandra Guha, Elizabeth Povinelli, and, more centrally still, Dipesh Chakrabarty, who has published a series of highly influential essays (most notably "The Climate of History: Four Theses," "Postcolonial Studies and the Challenge of Climate Change," "Climate and Capital: On Conjoined Histories," and "The Planet: An Emergent Humanist Category") arguing that climate change demands a fundamental reformulation of postcolonial studies' grounding interpretive protocols: its anti-universalism, its tendency to maintain a distinction between "human" and "natural" history, and its prioritization of cultural difference over humanity's collective "species" being.¹¹ In the epoch of the Anthropocene, addressing what it means for humanity to have become such a "geophysical" force is, Chakrabarty suggests, among the foremost tasks of critical thought and an inescapable frontier of postcolonial critique.

When I first read the first of these pieces, "The Climate of History: Four Theses," Chakrabarty's essay struck me with a force I can attribute to very few other pieces of contemporary theory I have encountered over the past quarter century—as much for the iconoclastic power of its argument as for the significance of the climate-historical dilemma it and the subsequent essays identify as a now inescapable condition of critical thought. I highlight this primarily to register that sense of his work's generative force and importance, particularly since, in the pages that follow, I will disagree with Chakrabarty as often as I will build on the challenges he maintains this Anthropocene "planetary conjuncture" puts to the core analytic strategies of the field and to my own previous attempts to consider the ways in which postcolonial studies might help us understand the making of the modern world.¹² If, heretofore, the questions I have addressed in my research have been questions firmly within the literary, legal, and archival realm of recorded human history—questions of the passage of bodies, laws, financial instruments, philosophical systems, narratives, political theories, and war machines in and around the port cities, capitals, coastlines, underwater burial grounds, and imperial littorals of the British empire and the Black Atlantic world—Chakrabarty's new work, and the rapidly emerging science of the Anthropocene on which it draws, suggests that if we are to continue to speak of the modern, we now need to do so with *natural* history (and, more crucially, the indistinction of human and natural history) also in mind.

This is far from a straightforward task. It is not simply a matter of addressing the political and cultural histories of modernity on the one hand and the modern world's natural history on the other. Speaking from within a moment in which the distinction between human and natural history has collapsed implies speaking from within a moment in which, as Chakrabarty has further argued, we are compelled to ask whether "the ideas about the human that usually sustain the discipline of history," and the methodological commitments of the humanities more generally, can survive that collapse—whether the philosophical discourse of modernity has anything to say to the discourse of the planet's anthropogenetically altered "natural" or *postnatural* history.¹³

The book that follows constitutes my attempt to answer that question, and to answer it strongly in the affirmative, to insist that what we know of the poetics of relation, of war zones and contact zones, of sovereignty, of bare life, raced life, and precarious life, of aesthetics, affects, and genres has a continued salience as we confront the crisis of climate change—even as what we are learning of climate change must re-affect our understanding of the enduring and future conditions of human history, culture, and experience across not only the Atlantic but all the world. Or, to reduce it to a formula, the argument that follows is that our understanding of the *force* of human politics, history, and culture must be held in interpretive tension and dialectical exchange with what we are discovering of the *forcings* of climate change as we address the fully planetary condition of the Anthropocene.[14] As the second volume in what I project to be a three-volume series on the place (and unfinished business) of the Black Atlantic in the making of the modern and postmodern world, this text seeks to put my earlier investigation of transatlantic slavery, speculative discourse, and finance capital in the making of the modern into relation with the overlapping place of carbon, climate, and environmental subalternity as these have also shaped the Atlantic—and, in shaping the Atlantic, have significantly shaped the modern in its now simultaneously "global" and "planetary" moment of arrival.[15]

To simply insist on the pertinence, significance, and power of our prior conceptions of history and interpretive protocols to the planetary conjuncture in which we find ourselves is, however, not enough. If we wish to find in the legacies and futures of those modes of understanding not only the continued relevance of the critical humanities to our times but the keys to rendering an engaged humanities a force contending with (perhaps even equal to) the planetary forcings of climate change we now confront, then—it is also my argument here—we will need to find a method for doing so, a method adequate to the situation of our multiscaled, period-multiplying Anthropocene times. Doing so means that we will also need to find a method capable of extending and reinventing a tradition of critical thought that has long understood its vocation as simultaneously descriptive and transformative: a method oriented to mapping the situation in which we find ourselves and to making something emancipatory of that situation; a method of thought committed, in the terms of Marx's "Theses on Feuerbach," not only to "interpret[ing] the world" but also "to chang[ing] it."[16]

What is the nature of that planetary world? What form does it take, not only, as I will repeatedly ask, as it comes to bear on the Ghanaian coast, on this one day, March 7, 2010, throwing into stark and ominous relief the image of this one child, Collins Kusietey standing in this one place in the dawning epoch of the Anthropocene, but the world in its planetary generality? What is this new world of the Anthropocene? What planetary conjuncture does it describe?

At its most technical sense, the answer is clear enough. First introduced by the atmospheric chemist and Nobel laureate Paul Crutzen and his colleague Eugene F. Stoermer in a 2000 newsletter of the International Geosphere-Biosphere Programme, the term *Anthropocene* is intended to identify a new epoch of geological time (following the Holocene) in which the core geological condition and future of the planet have been fundamentally reconstituted by human actions, anthropogenetic global warming (AGW) foremost among them. As Crutzen and Stoermer put it:

> Considering [the] . . . major and still growing impacts of human activities on earth and atmosphere, and at all, including global, scales, it seems to us more than appropriate to emphasize the central role of mankind in geology and ecology by proposing to use the term "Anthropocene" for the current geological epoch. The impacts of current human activities will continue over long periods. According to a study by Berger and Loutre, because of the anthropogenic emissions of CO_2 [carbon dioxide], climate may depart significantly from natural behaviour over the next 50,000 years. To assign a more specific date to the onset of the "Anthropocene" seems somewhat arbitrary, but we propose the latter part of the 18th century, although we are aware that alternative proposals can be made (some may even want to include the entire Holocene). However, we choose this date because, during the past two centuries, the global effects of human activities have become clearly noticeable. This is the period when data retrieved from glacial ice cores show the beginning of a growth in the atmospheric concentrations of several "greenhouse gases," in particular CO_2 and CH_4 [methane]. Such a starting date also coincides with James Watt's invention of the steam engine in 1784.[17]

Crutzen and Stoermer's proposition presented two fundamental questions for debate by the scientific communities charged with formally identifying the earth's periods of geological time: the International Union of Geological Sciences and the International Commission on Stratigraphy. Should this new epoch be officially recognized? And if so, from what point should it be dated? To grasp the magnitude of those questions, which the geologist Jan Zalasiewicz has called "arguably the most important . . . of our age—scientifically, socially and politically," we must first grasp where the epoch sits as a formal unit of measure within the continuum of "deep time."[18] That order of geological measure begins with *eons*, which encompass "hundreds of millions—or indeed billions—of years," and proceeds

> through smaller packages of time, such as the *eras* [lasting for scores of millions of years]. . . . These in turn are subdivided into *periods* of geological time, such as the Cambrian or Cretaceous. . . . Periods are divided further into *epochs* and *ages* . . . so well constrained that we can correlate such units globally and reconstruct the appearance and conditions of our planet for many hundreds of different time slices. The last period of time, the Quaternary, began just 2.6 Ma [million years ago], and includes two epochs, the Pleistocene and the Holocene. The latter—by far the shortest in the geological time scale—began only about 11,500 years ago, witnessed by changes in climate that manifest in an ice core from Greenland. . . . We distinguish it as an epoch for practical purposes, in that many of the surface bodies of sediment on which we live—the soils, river deposits, deltas, coastal plains and so on—were formed during this time.[19]

Within this orderly "slicing" of the immense expanses of the geological time scale, the significance of the question of the Anthropocene becomes apparent to Zalasiewicz et al. Have the "anthropogenic changes to the Earth's climate, land, oceans and biosphere," produced in a shockingly brief period of geological time, been so great that "an epoch-scale boundary has been crossed"? Their firm and grim conclusion is that that boundary has been crossed, not over the course of millennia, or millions of years, but "within the last two centuries," a mere geological instant whose numbingly accelerated pace of change, by the calculations of the University of Chicago geophysicist David Archer, has nevertheless succeeded in establishing a novel set of boundary conditions for the earth's climate that will endure for somewhere between the next 3,000 and 100,000 years.[20]

As Archer's work stresses, grasping the magnitude of the effect of post-eighteenth-century carbon dioxide and other greenhouse gas emissions on the geological future of the planet requires understanding the relative significance of these gases as one of four "agents of climate change called climate forcings that can warm or cool the climate." Two of these forcings are human-driven (anthropogenic): greenhouse gases and "sulfur from coal burning, which forms a haze in the atmosphere reflecting sunlight back to space to cool the Earth." The other two climate forcings are "natural": "volcanic eruptions and changes in the intensity of the Sun" caused either by sun flares ("long, slow flickers in the fires of the Sun" spanning decades or centuries) or by 10,000-year "wobbles" in the earth's orbital trajectory around the sun.[21] The relative impact of these four forcings can be compared by measuring their positive or negative effect on the average distribution of energy over a given area of the earth's surface, a measurement calculated "in terms of watts per square meter [w/m^2]."[22] Across the deep history of the planet, volcanic eruptions, whose sulfur emissions deflect sunlight and so cool the earth, have carried the greatest immediate capacity to effect climate change (at a rate of forcing of approximately negative 10 w/m^2), while solar flares (whose impact has generally been at the level of 0.1 w/m^2) have exerted the weakest event effect.[23]

Magisterially framing these episodic downward and upward bounces, however, have been the effects on the climate of cyclical 10,000-year-long wobbles in the earth's orbit around the sun. As Archer explains, these "orbital variations drive the climate by allowing ice sheets to grow or causing them to melt" and are thus responsible for establishing parametric conditions for the earth's climate by settling the planet into alternately glacial and interglacial multimillennial ages (we are now in an interglacial age). Across the deep history of the planet, the forcing effects of sun flares, volcanic eruptions, and human actions have always been framed (or, so to speak, provincialized) by these massive orbital forcings.[24] Archer's bleak warning is that over the past 250 years (and even more intensely over the past half century—the period of global warming often referred to as the "great acceleration") this has changed. We have now released so much carbon dioxide into the atmosphere (350 gigatons since 1750, with a total of 2,000 gigatons possible by the end of the twenty-first century at current rates of emission) that we risk not simply accelerating the melting of the polar ice caps but deferring the planet's anticipated cyclical transition into the next glacial age for anywhere between 50,000 and 130,000 years.

"Human climate forcing," in Archer's daunting summary, "has the potential to overwhelm the orbital climate forcing, taking control of the ice ages.

Mankind is becoming a force in climate comparable to the orbital variations that drive the glacial cycles."[25] "The practical implication," he concludes, "is that natural cooling driven by orbital variation is unlikely to save us from global warming.... [By] releasing CO_2, humankind has [acquired] the capacity to overpower the climate impact of Earth's orbit, taking the reins of the climate system that has operated on Earth for millions of years."[26] With such bleak findings in mind, it is perhaps no surprise that subsequent to the 2011 publication of the *Philosophical Transactions* special issue on the Anthropocene, the International Union of Geological Sciences (IUGS) and its official Subcommission on Quaternary Stratigraphy has formed a Working Group on the Anthropocene, convened by Zalasiewicz (the special issue editor), whose task is to forward a proposal to formally acknowledge the Anthropocene as a "defined geological unit within the Geological Time Scale."[27] At the time of my writing, the IUGS had not yet ruled, but as the final line of the Working Group on the Anthropocene's official website indicates: "It is widely agreed that the Earth is currently in this state."[28]

Thus the view from the geosciences.

But how does that enter into exchange with the view from the "human sciences"? How does it reframe the view from Totope?

When we first looked at this line of coast, we could not but see a multi-century process of the coming into being of the modern; a great play of historical forces; the African, European, and American agents of those forces; the complex co-constitution and dialectical interplay of sovereign power, capital, and disposable life; the subaltern refusal of disposability; the determination to survival; the creolization of culture; and the sweeping global necessity of a politics of freedom: in a word, the Black Atlantic and its defining place in the history of the modern. If we now regard this string of villages not only as dispersed minor capitols of the time-accumulating Black Atlantic but also as the outposts of the millennial coming into being of the Anthropocene, must we see the relation between time and power fundamentally differently? Should we radically elongate our scales of time and radically disperse our conceptions of power across a mingled human and natural spectrum? Or does this "both/and" formulation (this invitation to see Collins Kusietey's ocean-eaten home both as an outpost of the Black Atlantic and as a ruinous frontier of the Anthropocene) avoid the starker point?

If, indeed, we "cannot not periodize," then does seeing the village of Totope through the geo-optics of the Anthropocene so provincialize our existing periodizations of human history that we must set aside what we have known of this village and this coast's place in the history of Black Atlantic modernity and see them instead and solely within an epoch of geological time stretching from now into the almost unimaginable future? Must Totope take its place in an epoch in which, by virtue of the interplay between the planet's carbon cycle, the earth's multimillennial patterns of orbit around the sun, and the cumulative effects of human-generated (anthropogenic) warming, the distinction between human and natural history has not only collapsed but been so utterly swept away that all those *forces* of history that for so long have been the key focus of one or other order of critical materialist thought retreat in significance in comparison with the *forcing* power of climate change spreading over the planet, calving ice-sheets from the poles, melting glaciers, destabilizing the West Antarctic Ice Shelf, raising sea levels, and generally making the sovereignty of ice (resilient or dissolving), rather than the sovereignty of nations, empire, or capital, sovereign over this little seven-year-old boy, and the village, nation, Atlantic world, global commons, human species, and planetary ensemble he is made to stand in for?

What does it mean to see this snapshot of the Ghanaian coast through the lens of the Anthropocene? What do we recognize in the waters rising up behind Collins Kusietey's shoulder? The Black Atlantic or the Atlantic? The force of history or the forcings of climate change? An ocean-fronting world "still" made, in Christina Sharpe's resonant terms, in the "wake" of slavery: made and made and made again in the unending wake of the dungeon, wake of the factory, wake of the hold, wake of the plantation, wake of the door of no return?[29] Or do we see a world radically and singularly remade in the wake of carbon accumulating in the atmosphere? Do we see a postcolony unendingly subject to Sharpe's "pervasive," repeating, unending "weather" system and world "climate" of "antiblackness"?[30] Or do we see a planet subject to a radically new, anthropogenetically altered climate of life? Do we see the sovereignty of race-capital or the sovereignty of ice? An enduring call to the politics of human freedom and the task of critical thought in advancing that project or a new call to some other project of planetary politics? A human-produced system of enclosure, spreading from the Gold Coast's slave factories, across the waters to the New World plantations, and then, over the centuries, circling and re-circling back, through and around a circum-Atlantic archipelago of post-emancipation ghettos and postcolonial shantytowns? Or something else, something post-human, a new barracoon, rising up, millimeter by millimeter, yard by shore-encroaching yard, against which critical thought must now reorient itself and its unending spirit of emancipation and futures-shaping possibility?

Or are those false choices? Is it possible to see both these things at once? Is it possible to hold both the periods of human history and the epochs of geological time, the dynamics of forces and the operations of forcings, in concert and dialectical interplay with one another?

And if so, how so?

Those are the key questions of this book—and in a certain sense, as I have already indicated, their answer is quite direct.

Can we hold the view from the Anthropocene and the view from the Middle Passage in concert? Can we think simultaneously through historical periods and geological epochs, in time scales of centuries and across multimillennial spans? Can we think, in tandem, Sharpe's unchanged climate of antiblackness structuring the globe and Chakrabarty's new climate of history? Can we discern in the current conditions of the Ghanaian coast a new "poetics of relation" braiding together the factory production of the modern, an Atlantic cycle of capital accumulation, the orbital passages of the earth around the sun, the contemporary anthropogenic overwhelming of the planet's glacial cycles, and the slow but relentless rise in the level of the sea?

We can, and we must.

We can for the simple reason that the play of historical forces and climate forcings are not autonomous from one another but exacerbate and intensify one another. We can because, over the centuries, the forces requisite to the slave trade, the forces of modern power, and the forces of global capital concentrated in the Ghanaian factories have continuously gone to work on the forcings of the carbon cycle. In their encounter with and acceleration of an age of fossil fuels, those forces have helped create and intensify the global practices of consumption and the cycles of carbon accumulation that have thrown us into the planetary moment of the Anthropocene. But there is more than this. The intimate interrelationship also works the other way around. While the forcings of climate change are, by one order of measurement, smoothly, evenly, spherically distributed across the planet, they are also asymmetrical in their impact on the globe; both indifferent and highly uneven in their distribution of vulnerabilities; implacably unaware of and accelerating the material residues and impacts of prior and current human historical forces, including those forces of history that continue to structure the composition of life along the Ghanaian coast and to make of it not only a geological but a *human* shore.

Can we frame a dialectic of forces and forcings? We can, and we must.

But to insist so requires naming four key things. The first is the necessity of exploring the relation between a longstanding tradition of historical materialist thought (what I will refer to as *Materialism I*) oriented toward addressing the sort of historical forces animating the operations of the coastal, Black Atlantic factory system I have sketched, and an emerging, newer materialism (which I will call *Materialism II*) specifically attuned to numbering a post-human realm of carbon, sun flares, wattage, glaciers, heat waves, ocean deserts, orbital wobbles, and radiative forcings ("hyper-objects," according to Timothy Morton, or "vibrant matter," for Jane Bennett) as among humanity's Anthropocene fields of "circumstance"—as, in Bruno Latour's, Morton's, Donna Haraway's, Achille Mbembe's, and Pope Francis's terms, the planetary co-"actants," "strange strangers," "cosmological assemblages," and "companion species" of our "common home."[31]

The second element to approach carefully is the question of method, moving beyond the assertion that we *can* hold these two materialisms in concert to a detailed inquiry on *how* we might do so. In developing that inquiry on method, one of my chief purposes will be to underscore that this relation is not nearly as obvious, smoothly bidirectional, or matter-of-fact as my previous comments might imply. As Chakrabarty has made manifest in his discussions of this exact issue, the magnitude of the methodological challenge climate change puts to many of the long-standing disciplinary assumptions of the humanities cannot be overstated. Surveying the enormity of planetary transformation that climate change is producing, he has offered this stunning admission: "All my readings in theories of globalization, Marxist analysis of capital, subaltern studies, and postcolonial criticism over the last twenty-five years, while enormously useful in studying globalization, had not really prepared me for making sense of this planetary conjuncture within which humanity finds itself today."[32] This warrants a pause, and I will return to the implications of this statement throughout much of my argument. While—as I have already indicated—my ultimate conclusion tends in a different direction (by claiming that globalization theory, subaltern studies, postcolonial criticism, Black Studies and Black Atlantic Studies, and the range of other critical materialist practices that fall under the rubric of Materialism I remain vital to making sense of our current planetary conjuncture), I am in complete agreement that those practices of critique cannot do so alone. To borrow another of Chakrabarty's formulations, these grounding practices of historicist thought (Materialism I) are to the emerging epistemologies and ontologies of the Anthropocene (Materialism II) as Enlightenment thought itself is to postcolonial theory: "indispensable" *and* "inadequate."[33] We must therefore take seriously the methodological challenge Chakrabarty highlights. Detailing the relationship between a particular tradition of subaltern and postcolonial critique and the new materialist epistemologies associated with the work of Latour, Bennett, Morton, and others in their reciprocal insufficiency, mutual necessity, and complementary interdetermination as paradigms for responding to the challenges of the Anthropocene is the second key labor of this book.

The third is to consider how the relation of these two materialisms invites us not only to generate new forms of critical method but to consider the possibility that their coming together reveals the emergence of a new order of Anthropocene time: an order of time knotting together what I will call the historical, the infra-historical, and the supra-historical; an order of time whose comprehension requires a braiding together of the epistemologies and ontologies of a classically progressive historicism, a counter-historicist subaltern conception of "now-being," and a post-humanist articulation of object-oriented time; a deeply hybrid order of time I understand the globe/planet now to be entering under the advent of the Anthropocene.

The fourth key question is not only whether this dialectic of forces and forcings can be imagined, or how it can best be framed, but *why* it must be articulated. From the particular positional grounds of postcolonial, diaspora, and Black Atlantic studies, that question translates in this way: why should a range of knowledge fields long attuned to aligning their practices of understanding to advancing those global projects of freedom that, in Chakrabarty's terms, have provided "the most important motif of written accounts of human history of [the past] two hundred and fifty years," now turn attention from the subordinating forces of race and capital and empire to Archer's planetary "forcings"?[34]

The answer is as evident as it is necessary to state. It is because these new forcings, as they impact the conditions of life on the planet, manifest themselves also as forces of profound violence and unfreedom; as forcing-forces for the reactivation of old and the animation of new modes of subalternality, inequality, and vulnerability; as the magnifiers and accelerators, in Rob Nixon's terms, of an extraordinarily extensive process of "slow violence" visiting its devastation on the twenty-first century's "wretched of the eaarth."[35]

To put it another way: "Men make their own history," as Marx observed in a famous passage from *The Eighteenth Brumaire of Louis Bonaparte*, "but they do not make it just as they please, they do not make it under circumstances chosen by themselves, but under circumstances directly found, given, and transmitted by the past."[36] We are long accustomed to understanding how complex, and how disposed to human constraint, is that field of "circumstances . . . transmitted by the past." Much of the work of the humanities and interpretive social sciences for the past decades (and more) has been to identify the range of forces creating these circumstances of unfreedom (forces of racialization, of normalization, of gendering, of dispossession, of biopoliticization) and of directing the opposing force of critique against them. The epoch of the Anthropocene does not leave that work behind. Indeed, one of the key epistemological and methodological imperatives of philosophical critique in the epoch of the Anthropocene is to link our prior investigations of such forces to an examination of these newly visible climate forcings, to understand how the prior and enduring conditions of unfreedom these prior and enduring forces have layered around human life are now being exacerbated and intensified, slowly and explosively, by the forcings of the Anthropocene.

As Ato Quayson has observed, crossing over to the geological epoch of the Anthropocene does not mean crossing into a historical or geopolitical tabula rasa. To say that the human has been "converted into a geophysical force" does not mean that humanity ceases to be, simultaneously, a "sociopolitical category" structured by a multitude of historical and anthropological differences, nor does it mean that we should now adopt a hermeneutic of planetary "equivalence" that would treat "the climate change displacing hundreds of thousands in today's southern Sudan" as purely identical with "the climate change that led to Hurricane Katrina and the terrible displacements that ensued in New Orleans in 2005." "Man-as-geophysical force," Quayson entirely correctly concludes, is "in each instance . . . the product of specific political, social, cultural, and economic realities."[37] To attend to the forcings of the Anthropocene is, therefore, crucially, also to attend to the legacy and persistence of those political, social, cultural, and economic forces that unevenly structure human life on the globe and unevenly distribute the lived effects and vulnerabilities of climate across the planet. It is, in short, to attend to the complex dialectic of forces and forcings. It is also, reciprocally, to grasp the intensifying power of these forcings as historical forces, to address carbon as a forcing-force with the force-power to disrupt and devastate polities, economies, societies, cultures.

How significant is the combined geological and geopolitical power of those CO_2 forcings? How fully are they reshaping the "circumstances" under which history can be made and experienced across the planet-world and in the contemporary postcolony?

There is no unitary answer, but according to the projections of the working group of the Fifth Assessment Report (AR5) of the United Nations Intergovernmental Panel on Climate Change (IPCC), there are now four distinct "climate futures" or routes toward which we may be hurtling. By the terms of the report, each "Representative Concentration Pathway" (RCP) predicts "scenarios" for the future according to anticipated average "radiative forcing values" for the planet in the year 2100.[38] Each of those pathways is calculated in units of measure with which we are now familiar, w/m² (watts per meter squared), and are named by those measures. Thus the IPCC's four climate futures are identified as RCP2.6 (a future scenario with a planetary 2.6 w/m² increase in radiative forcing) and, respectively, RCP4.5, RCP6, and RCP8.5. Subtending each of these pathways are a bundle of climate "drivers" effecting radiative forcing, including estimates of global population growth, patterns of future land use, changes in global GDP, technological development, and, most crucially, "increase[s] in the atmospheric concentration of CO_2."[39] As might be anticipated, the mildest pathway, resulting from the lowest increase in radiative forcing (RCP2.6), would require the virtually immediate adoption of "stringent climate policies to limit emissions."[40] Representative Concentration Pathway 4.5, correspondingly, would result from the implementation of more modest "climate policy scenarios." RCP6 and RCP8.5, by contrast, represent "forcing pathway[s]" into the future in the general or wholesale absence of global climate policies and significant changes in patterns of world carbon emission. To each of these avenues to the future, the IPCC report then assigns a predicted change in "global mean surface temperature," calculated in degrees Celsius (by the year 2100).

For RCP2.6 that anticipated change by the end of the current century would be between 0.3°C and 1.7°C.

For RCP4.5 it would be between 1.1°C and 2.6°C.

For RCP6 it would be between 1.4°C and 3.1°C.

For RCP8.5 it would be between 2.6°C and 4.8°C.[41]

None of those scenarios are good. As the report makes clear, each pathway will generate "changes in all components of the climate system," altering the "global water cycle," negatively impacting air quality, exacerbating ocean warming, decreasing Arctic sea ice and "global glacier volume," accelerating the rise of sea levels, and increasing "ocean acidification."[42] The truly catastrophic results of climate change, however, attend the last two pathways, RCP6 and RCP8.5, as temperature change in these scenarios approaches or crosses the four degrees Celsius (4°C) line that gives the title to this volume. The initial document of the Fifth Assessment Report does not spell out the consequence of that order of temperature rise, but a report issued a few months earlier by the World Bank does. The full title of the report is "4°—Turn Down the Heat: Why a 4°C Warmer World Must Be Avoided."[43] It is one of many similar documents produced between Climate Change 2007, the IPCC's Fourth Assessment Report, the Fifth Assessment Report, and the landmark December 2015 Paris climate accord, the key purpose of which was not so much to prevent climate change as to desperately attempt to keep the planet within the 2°C threshold of heating predicted by RCP2.6 rather than crossing into RCP8.5 and a 4°C change above average pre–industrial era temperature levels. Like virtually all of the documents produced within this period of climate discourse, the World Bank report makes for chastening reading, if only for the simple clarity of the danger threshold it identifies: the 4°C world we must prevent but, which, if current emission trends continue unabated, will become our world "within this century."[44]

What does it look like, that 4°C world the future may well inherit from its carbon-era present and carbon-era past (that past, which, in this sense, is not so much past as it is accumulating, moment by moment, one carbon part per million in the atmosphere)?

Whatever else it will be, the World Bank report indicates a 4°C world will be changed comprehensively and disastrously across almost every sector of analysis. It will be a world in which "extreme weather events" will intensify both in frequency and in scale, with "heat waves such as [the one] in Russia in 2010 [which killed an estimated 55,000 people] likely to become the new normal summer."[45] It will be a world in which, for "regions such as the Mediterranean, North Africa, the Middle East, and the Tibetan Plateau, almost all summer months are likely to be warmer than the most extreme heat waves currently experienced."[46] It will be a world in which, as warming "strengthens the [planetary] hydrologic cycle . . . [and] dry regions . . . become drier and wet regions . . . wetter," there will be increased mass flooding in some regions of the globe (much of the Northern Hemisphere, East Africa, and South and Southeast Asia), and the simultaneous sprawl of aridity and desertification in other zones, leading to "dramatic reductions in global agricultural production," with "35 percent of [all sub-Saharan African] cropland . . . expected to become unsuitable for cultivation."[47]

It will be a world in which melting Greenland, Antarctic, and Arctic sea ice "will likely lead to a sea level rise of 0.5 to 1 meter, and possibly more, by 2100, with several meters [and possibly significantly] more to be realized in the coming centuries."[48] It will be a world in which coastal communities around the world, and a "highly vulnerable" archipelago of cities in Mozambique, Madagascar, Mexico, Venezuela, India, Bangladesh, Indonesia, the Philippines, and Vietnam will, in consequence, find themselves exposed to "extreme floods" and "coastal inundation."[49] It will be a world in which such "large scale extreme" flooding "events" will drown people, collapse buildings, "induce nutritional deficits" due to the loss of arable land, and increase "diarrheal and respiratory diseases" by introducing "contaminants and diseases into healthy water supplies."[50] It will be a world in which, even as such coastal flooding exerts massive impacts on human health, compounding chronic "changes in temperature, precipitation rates, and humidity [will further] influence vector-borne diseases (malaria and dengue fever) as well as hantaviruses, leishmaniasis, Lyme disease, and schistosomiasis," and exacerbate respiratory disorders and heart and blood vessel diseases" due to "heat-amplified levels of smog."[51]

Farther out to sea, the oceans will intensify their rate of acidification, leading to a significant loss in biodiversity in the Atlantic, Pacific, and Indian Oceans' marine ecosystems; correspondingly dramatic reductions of fishery yields; and the widespread dissolving of coral reefs, with "profound consequence for [the reefs'] dependent species and for the people who depend on them for food, income, tourism, and shoreline protection."[52]

And these, the World Bank report indicates, are just some of the likely *linear* effects of warming. "Lurking in the tails of the probability distributions," it more ominously warns, "are likely to be many unpleasant surprises . . . [as] extremes, including heat waves, droughts, flooding events, and tropical cyclone intensity, are expected to respond *nonlinearly* . . . [leading to an] evolving cascade of risks," including "largescale displacements of populations, with manifold consequences for human security, health, and economic and trade systems"; "the risk of crossing activation thresholds for nonlinear tipping elements in the Earth System"; and, as the report notes in a concluding gesture toward just how much the damage might exceed its probability calculations and risk scenarios, "the likelihood of transitions to unprecedented climate regimes."[53]

What will a 4°C world look like? A world of the long catastrophe; a world that finds itself, at best, entering the moment of "ultra-history," in Giorgio Agamben's term; a world in the long interregnum between the accumulating certainty of the devastated and the uncertainty of the new.[54] In the language of the eschatological and apocalyptic tradition on which Slavoj Žižek has drawn, it looks like a world of "the End Times": a world possessed of a "new heaven and a new earth."[55] Or as Bill McKibben's evocative neologism has it, it will be the world of a new "Eaarth": a world radically different from what we have heretofore understood the planet to be.[56]

Why should a knowledge field long attuned to advancing those global projects of freedom that have provided "the most important motif of written accounts of human history of [the past] two hundred and fifty years" now add to its analysis of the subordinating forces of race and capital and empire an encounter with Representative Concentration Pathways and planetary climate forcings?[57]

Why should a discourse on the Black Atlantic now also become a discourse on the Atlantic Ocean?

Because, to return to the place where I began, the slave factories on the Black Atlantic coast of Ghana and the village of Totope that was once (but is no longer) three miles from the sea do not belong to separate worlds but to overlapping worlds. Because, if we have known, or thought we have known, what challenges to freedom the slave factories presented, we now need also to ask what challenges to freedom those rising waters threaten.

Or let me put it this way:

Achille Mbembe opens the introductory chapter ("The Becoming Black of the World") of his magisterial *Critique of Black Reason* thus: "I envision this book as a river with many tributaries, since history and all things flow toward us now. Europe is no longer the center of gravity of the world. This is the significant event, the fundamental experience, of our era."[58] The "us" in Mbembe's epoch-claiming statement is a black "us," a black "we" whose history, condition, and future has now become the history, condition, and future of the world (or, as he also indicates, the "planet").[59] By his account, three "critical moments in the biography of the vertiginous assemblage that is Blackness" have led to this epochal becoming-black of the world.[60] The "first arrived with the organized despoliation of the Atlantic slave trade (from the fifteenth through the nineteenth century) through which men and women from Africa were transformed into human-objects, human-commodities, human-money."[61] The "second moment corresponded with the birth of writing near the end of the eighteenth century, when Blacks, as beings-taken-by-others, began leaving traces in a language all of their own and at the same time demanded the status of full subjects in the world of the living."[62] The "third moment—the early twenty-first century—is one marked by the globalization of markets, the privatization of the world under the aegis of neoliberalism, and the increasing imbrication of the financial markets, the postimperial military complex, and electronic and digital technologies."[63]

The capture and thingification of black life; the revolt and rehumanization of black life; the reactive financialization and equitization of black life: these, for Mbembe, are the three epochs in the history of Blackness. And they are equally and at the same time more than that. They are, in his counter-Hegelian response to Hegel, the three epochs in the history of the modern world.[64] They are the three serially accumulating epochs of a contemporary world era in which all human life on the planet falls available for subaltern capture; all human life finds itself called and stirred by that incomplete "promise of liberty and universal equality" that has been the beacon of black political struggle "throughout the modern period";[65] and all human life—planetwide, whether captive, in revolt, or putatively "free"—lies endlessly susceptible, endlessly exposed, to becoming a tradeable/abandonable/superfluous good and quantum of metrical-virtual value or threat, an "animate thing made up of coded digital data."[66] As Mbembe summarizes his argument: "Across early capitalism, the term 'Black' referred only to the condition imposed on peoples of African origin (different forms of depredation, dispossession of all power of self-determination, and, most of all, dispossession of the future and of time, the two matrices of the possible). Now, for the first time in human history, the term 'Black' has been generalized. This new fungibility, this solubility, institutionalized as a new norm of existence and expanded to the entire planet, is what I call the Becoming Black of the world."[67]

My arguments in what follows are in agreement, particularly with the corresponding point on which Mbembe insists: as critical thought in the conjoined eras of late capital and the Anthropocene (or as Jason W. Moore and others have indicated, in the age of the "Capitalocene") turns from the category of the human to the category of species, that does not entail leaving behind the history of race—neither the forces of history through which race has been subalternized, captured, dungeoned, traded, or financialized, nor the forces of liberatory revolt and struggle for freedom against all these closures, enclosures, distributions, trades, and financializations, what, in summary, Sharpe calls the force of those new "ecologies" of freedom "produce[d] out of the weather" of black being—but that, instead, we must read the contemporary discourse on species as raced; must read the turn to "species" as an extension and planetary expansion of the history of race and blackness; must read "species" as a critical instance of the vulnerable, disposable, subaltern, precarious "becoming black" of human life across the planet.[68] As Mbembe efficiently puts it, "We [must] understand that as humanity becomes fungible, racism will simply reconstitute itself in the interstices of a new language on 'species.'"[69]

On these lines, Mbembe and Gilroy, despite all their other differences, are in agreement, particularly as Gilroy, in his 2014 Tanner Lectures on Human Values (written to mark the twentieth anniversary of the publication of *The Black Atlantic*) insists that the Black Atlantic archive remains utterly central to the task of articulating a trans-planetary politics of freedom adequate and responsive to the contemporary discourse on species: a politics, in his terms, that depends on the "elaboration of a planetary humanism" flowing from the histories, situations, and practices of Black Atlantic art, struggle, and thought.[70] Mbembe, of course, is far less sanguine regarding the appeal to a planetary "humanism," and I will be returning to this point of difference in what follows. For the moment, however, I wish merely to mark the significance that both Mbembe and Gilroy place on the continued urgency of a black radical tradition to our not merely "global" but "planetary" moment (or, as Chakrabarty has it, to that "planetary conjuncture within which we find ourselves"). Beyond this, Gilroy and Mbembe share something else. They share the conviction that the necessity of such a planetary encounter with the black force of history must now also encompass an encounter with the forcings of the Anthropocene; an encounter with a mode of vulnerability/disposability appearing not only under the generalized "aegis of neoliberalism" but under the crash of the climate-and-capital-changed oceans, coasts, cities, and political orderings of a new Anthropocene nomos of the Eaarth; an encounter with a new mode of precariousness to which black life is, at once, singularly subject and of which black life (or, to be more precise to Mbembe, black "forms of life") is/are prophetic of the "generalized" "becoming-black," planetwide, of the "species."

Gilroy is entirely direct in drawing this connection between the ongoing, on-flowing currents of Black Atlantic history and the rising, surging lap of the Atlantic Ocean and the world's other climate-altered seas. As he indicates in the opening sentences of his first Tanner Lecture ("Suffering and Infrahumanity"):

> The invitation to deliver these lectures coincided with the twentieth birthday of the publication of my book *The Black Atlantic*. That anniversary provided me with a cue to reflect critically on its reception, reach, and travel as well as to return to and develop a number of its key themes. Of course, the book's intervention resonates differently now that the "grey vault" of the sea is rising and smaller boats sweep fleeing Africans northward to fortified Europe rather than westward into the colonial nomos of plantation slavery.[71]

Despite those changes in circumstance, he continues: "[the] radical tradition of the black Atlantic," outlined by the earlier book, "remain[s] . . . an incendiary object. . . . [I]t is still endowed with the capacity to articulate conceptions of freedom, autonomy, and resistance that, though they derive from the struggle against racial slavery, remain not only intelligible but in some undefined ways also risky and relevant, even dangerous."[72] Holding that tradition together, he concludes, "is the overriding ethical and political task that can be said to distinguish the black Atlantic tradition, namely, the fashioning of a humanism made, as Aime Cesaire put it in the final sentence of his *Discourse on Colonialism*, 'to the measure of the world.' This is the task that I have described elsewhere as the elaboration of a planetary humanism. . . . We will need all its appeal as the sea levels rise and the fortifications placed around the citadels of overdevelopment crack open, releasing the pressure for new collectivities and solidarities as well as new modes of accountability to one another."[73]

While Mbembe, as I have already noted, would hesitate over that humanist inflection of a Black Atlantic tradition, he is nevertheless as convinced as Gilroy that any liberatory politics of the present must be routed through the force of black reason (where reason must be read, overlappingly, as thought/epistemology/purpose/aspiration/form/structure of time). Like Gilroy, he is further convinced that any such politics of the present (and its futures) cannot conceive of our present as an abstract contemporaneity but must grasp it as an epoch structured by an entangled congeries of "processes."[74] I have already mentioned three of those processes (as Mbembe outlines them in the introduction to *Critique of Black Reason*), all of which, notably, concern the planetary generalization (from "blackness" to "the species") of situations of human life: the capture and thingification of life; the revolt and rehumanization of life; the financialization and equitization of life.

To these three constitutive orderings of the human, Mbembe adds another in his essay "Decolonizing Knowledge and the Question of the Archive":

> [W]e can no longer think about "the human" in the same terms we were used to until quite recently. . . . an epoch-scale boundary has been crossed within the last two centuries of human life on Earth . . . we have, as a consequence, entered an entirely new deep, geological time, that of the Anthropocene. . . . The scale, magnitude and significance of this environmental change—in other words the future evolution of the biosphere and of Earth's environmental life support systems particularly in the context of the Earth's geological history—this is arguably the most important question facing . . . humanity since at stake is the very possibility of its extinction. We therefore have to rethink the human not from the perspective of its mastery of the Creation as we used to, but from the perspective of its finitude and its possible extinction. . . . This rethinking of the human has unfolded along several lines and has yielded a number of preliminary conclusions I would like to summarize. The first is that humans are part of a very long, deep history that is not simply theirs; that history is vastly older than the very existence of the human race which, in fact, is very recent. And they share this deep history with various forms of other living entities and species. Our history is therefore one of entanglement with multiple other species. And this being the case, the dualistic partitions of minds from bodies, meaning and matter or nature from culture can no longer hold.[75]

Two points are worth registering here. The first is that as Mbembe marks this Anthropocene reordering of the human, he takes a step further than Gilroy in articulating a renovated philosophy of history that recognition demands. While never turning from the centrality of the histories of race (and the force of those histories) to our multi-entangled present, he nevertheless indicates that while the history of race is indissociable from the history of "the human race," human history can no longer disentangle itself from a deeper history of "other" forms of "entities and species."

It is at such moments that we can begin to see Mbembe and Gilroy parting ways on the need to link a contemporary black politics of the planetary to an a priori commitment to humanism. But that "deep time" and multispecies troubling of the humanist boundary (while certainly central to my own arguments) is not the most remarkable thing to note in Mbembe's comments. The second and more original of his arguments is that the Anthropocene-trespassed human/nonhuman boundaries he registers in this essay are not at odds but consistent with the digitally trespassed human/nonhuman boundaries he delineates in *Critique of Black Reason*. In both texts, he suggests, the central dilemma of "life" in the epoch of planetary entanglement is the dilemma of human life animated by and along with a vast bundle of nonhuman counteragents: the dilemma, as he puts it, of "animism," which has also been the ontological question and possibility on which a sustained, radical, and non-humanist reason of black thought has long nourished itself in its own encounters and contestations with the traditions of Enlightenment reason (most notably, the Enlightenment insistence on "the dualistic partitions of minds from bodies, meaning [from] matter, nature from culture").

Entangled with the screen, entangled with nonhuman biotic forms of life, entangled with data, entangled with surging oceans, entangled with equity bundles, entangled with the geological, entangled with algorithms, entangled with gene-coding, entangled with sun flares, entangled with derivatives—the human in the epoch of the planetary contemporary, Mbembe indicates, can no longer be imagined to hold its humanist core. Things fall apart—and together. The Eurocentric order of Enlightenment reason cannot hold: "[C]apitalism and animism—long and painstakingly kept apart from each other—have finally tended to merge."[76] A black ontology of the entanglement of human and nonhuman life, things, and matters—precarious and possible, abject and hopeful, dismissed and liberatory—becomes the ontology of the human species in its generalized totality—"a new norm of existence . . . expanded to the entire planet."

History and all things flow toward us now. Europe is no longer the center of gravity of the world. This is the significant event, the fundamental experience, of our era.

Thus, Mbembe's answer to Gilroy's call for a planetary humanism:

~~~ In place of a planetary humanism (nourished on the Black Atlantic tradition), a planetary animism (nourished by the black reason of that tradition of struggle, practice, and thought).

Or, to reduce things to one of the core arguments I will be developing:

~~~ Thus the difference between what I will be calling a (humanist) Materialism I and a (post-humanist) Materialism II in their responses to the dilemmas of our time.

However, as I have already indicated, and will be insisting on throughout, my ultimate purpose is not to hold these humanist and post-humanist responses to the crisis of our planetary time in irreconcilable tension with one another but to discover the points of their intersecting swerve. Finding those points of convergence, while attending to the significant differences of these approaches, will take up a substantial portion of this book. But one thing, one point of convergence, is already abundantly clear. Gilroy and Mbembe swerve together where Quarmyne's art also meets and extends what they hold to be the defining feature of the Black Atlantic tradition: they meet at the point of (and at the struggle for) freedom. Which is not to say that freedom will look the same in its humanist and post-humanist guises. But it is to say that all three of their bodies of work (as, also, this book) are sustained and driven by an investigation of the nature of freedom in our planetary times. And it is to say that I am in full agreement that in that search for such a conception of freedom adequate to our times (and in the corresponding search for a method of inquiry adequate to the task of the critical humanities in the epoch of the Anthropocene), Gilroy's Black Atlantic tradition and Mbembe's "becoming black of the world" articulate themselves as organizational starting grounds. As Gilroy, to repeat, says, "We will need all its [the Black Atlantic archive's] appeal as the sea levels rise and the fortifications placed around the citadels of overdevelopment crack open, releasing the pressure for new collectivities and solidarities as well as new modes of accountability to one another."[77]

As Mbembe—convergently and differentially—puts it:

> The term "Black" was the product of a social and technological machine tightly linked to the emergence and globalization of capital. It was invented to signify exclusion, brutalization, and degradation, to point to a limit constantly conjured and abhorred. . . . But there is also

a manifest dualism to blackness. In a spectacular reversal, it becomes the symbol of a conscious desire for life, a force springing forth, buoyant and plastic, fully engaged in the act of creation and capable of living in the midst of several times and several histories at once.... [Must we then] forget Blackness? Or perhaps, on the contrary, must we hold onto its false power, its luminous fluid, and crystalline character—that strange subject, slippery, serial, and plastic, always masked, firmly camped on both sides of the mirror, constantly skirting the edge of the frame? And if, by chance, in the midst of all this torment, Blackness survives all those who invented it, and if all of subaltern humanity becomes Black in a reversal to which only history knows the secret, what risks would a Becoming-Black-of-the World pose to the promise of liberty and universal equality for which the term "Black" has stood throughout the modern period?[78]

What risks does the Becoming-Black-of-the-World pose to antecedent Enlightenment notions of liberty and equality? What reinventions of freedom do Mbembe's "Black Reason" and Gilroy's "Black Atlantic tradition" demand? What new collectivities, solidarities, and modes of accountability do they variously and convergently invite? What promise can they offer to a seven-year-old boy standing on the fringe of one of the Black Atlantic's ocean-devoured shores?

The ensuing portions of this book constitute my attempt to take up these questions and concerns, building on the work of scholars across the disciplines who have begun to outline the terms for a method of critical thought adequate to this threshold-crossing moment of humanity's transit into a 4°C world—and to the call of liberatory thought in the midst of that epochal passage.

In the section of the book that follows, I provisionally put on hold a detailed reengagement with the contemporary West African scene that Quarmyne's images bring to light in order to trace the broad outlines of such a "method" for apprehending this alternatured planetary world. While the methodological reflections of this section flow from the intention to return to my site of departure—to that March morning of 2010 and the image of Collins Kusietey standing in his Anthropocene-altered slave-coast shore—they are also explicitly more general in scope, driven by the challenge of finding a method adequate to Gilroy and Mbembe's planetary moment, or, in McKibben and Chakrabarty's blended terms, to the Eaarth-spanning "planetary conjuncture" in which we find ourselves. The section thus begins by establishing a broader historical framework for the recent round of critical work directly engaging the theoretical and philosophical challenges of the Anthropocene. It does so by addressing a prior moment of philosophical debate, one that does not in any simple sense produce our theoretical contemporaneity but, which, nevertheless, played a crucial role in reopening for late twentieth- and early twenty-first-century debate a series of eighteenth- and nineteenth-century questions on the relation between freedom and finitude, the actor and the situation, and the interchanges and interminglings of human and extra-human history. In doing so, this moment helped articulate what I understand to be the key set of questions at the heart of our current theoretical and methodological problematic.

The moment I have in mind is that of the early 1960s—one generally characterized in histories of theory as marking the beginning of the transition from structuralist to poststructuralist epistemologies but highlighted for my purposes by Claude Levi-Strauss's celebrated 1962 dispute with Jean Paul Sartre on the adequacy of historical method to a properly dialectical understanding of the human "situation." This is a moment, consequently, of less interest to me for what it reveals regarding structuralism's contention with, and triumph over, existentialism (in advance of its subsequent poststructural defeat), than for the ways in which, by reviving a debate on Marx's *Eighteenth Brumaire*, and in throwing open the question of how to gauge the pertinence of a Marxist dialectic of freedom and circumstance to humanity's nonhuman fields of circumstantial determination, Sartre and Lévi-Strauss's heated quarrel helped open for investigation many of the animating questions of the various post-humanist epistemologies that have emerged in the ensuing years—particularly the "new materialisms" that have risen to such prominence in the early decades of the twenty-first century.

While not figuring themselves in relation to the Sartre/Lévi-Strauss debate, these new materialisms share and extend some of its crucial features, taking something (however paradoxically) from both sides. From the Sartrean-Marxist side of the debate, in particular, they take the call for a search for critical method adequate to indicating, in Sartre's terms, how we may "succeed in making *of* what [we have] been made."[79] From Lévi-Strauss's side of the debate, they take the argument that the "human" situation is inadequately addressed by the "historian's code," that the "circumstance," or situation, of our times must be understood to encompass multiple scales and orders of time radically exceeding what a classically historical (and residually humanist) materialism (Marxist or Sartrean) renders available for investigation and critique: most significantly, an array of extra-historical, infra-historical, and supra-historical orders of human/nonhuman being and time.

Having sketched this broader historical framework for our current critical engagement with the dilemmas of the Anthropocene, I return to Chakrabarty to discuss the ways in which his earlier writings and his recent work on climate change further extend the debate between these historical and extra-historical materialisms while adding a particular, postcolonial inflection to that debate's terms of inquiry. I pursue this reading by considering the relation between his earlier conceptualization (in *Provincializing Europe: Postcolonial Thought and Historical Difference*) of two forms of history (an Enlightenment form he calls History 1 and a subaltern form he calls History 2) and the new dilemma of history emerging from the bundle of climate change essays he has published since 2009. Despite its enormously rich considerations of the methodological challenges the Anthropocene poses to the "historian's code," Chakrabarty's new mode of understanding, I suggest, nevertheless bypasses the full multiscale temporality of our planetary present. I argue, therefore, that while drawing on Chakrabarty's recent work (and the moments of Enlightenment and Marxist/subaltern thought preceding it), we need to continue in a search for method adequate to the heterochronic situation of our Anthropocene time.

In making that argument, I am guided by a series of framing questions. If it is the case that a rapidly arriving Anthropocene world is temporally multiple, and that to its multiple temporalities there is a correspondingly multiple and heterogeneous set of orientations to the future, then what are the strands of time of which that world is composed? How, further, might we understand those temporal orders to relate to one another? Where might we look for models of their relationship? What projects of future-fashioning do they variously imply? And how, as they orient us to the future, do they reopen or renovate the question of freedom?

I respond to these questions by returning to Sartre to consider what his reflections on "totalizing" method (particularly in *Critique of Dialectical Reason*) may still have to teach us as we attempt to consider the emergingly "total" Anthropocene condition of the planet, and where they fall short in addressing a "situation" that radically exceeds reduction to a persistently humanist "historian's code." The Anthropocene, to put it another way, may register the present and coming *total* history of the planet, and Sartre's method may provide an avenue to grasping such a totality, but his persistent tendency to render the "situation" of dialectical critique exclusively isomorphic with human history also underscores the limits of an older historical materialism's capacity to address not only the historical forces of climate change but also the supra-historical forcings of the Anthropocene. To consider how we might more fully address the situation-altering and situation-expanding effects of those forcings, I turn to Walter Benjamin, particularly to the relatively underexamined eighteenth (and final) thesis in his "Theses on the Philosophy of History," which dramatically scales up his prior account of the complex *now-being* (*jetztzeit*) of historical time to include a range of evolutionary and geological orders among humanity's temporal fields of circumstance.

Working from that text, I argue that Benjamin provides us not only with a model for a renovated philosophy of history but with a method for grasping an internally heterogeneous "totality" structured by a series of biographical, biological, nomological, geological, cosmological, and theological scales of being and time. With this fuller sense of how to fashion a method for grasping the plural temporalities and mixed ontology of the Anthropocene more firmly in place, I conclude this section on method by addressing the ways in which aesthetic experience—particularly the aesthetic experience of encountering the new forms of "character" emerging in recent climate-change novels and visual art—vitally supplements philosophical critique in helping reveal what it means for human life to be distributed across this range of temporal scales and ontological registers; what it means for the human to be, simultaneously, a bearer of rights, a subject of cultural difference, an expression of co-evolutionary deep time, a geophysical force, and a measure of the infinite; what it means to pose the question of freedom from within these multiple "situations" of human and nonhuman life—serially, and all at once.

In the third section of the book, I return to directly reengage the text's Atlantic scene of departure. Having outlined a general method of approach, I begin the section by asking how that method comes to bear on the twenty-first-century Ghanaian shore, and how, reciprocally, thinking from this shore shows us how the discourses of the Anthropocene Atlantic and the "historical" Black Atlantic enter into exchange with one other (through that dialectic of forcings and forces I have been tracing). I take that question up through the question of form. Identifying a series of temporal scales ordering the epochal time of the Anthropocene (a biographical scale, a nomological scale, a biological scale, a zoëlogical scale, a geological scale, and a cosmological scale), I then ask how Quarmyne's work might help us more fully understand the form of these time scales' relation to one another. If the Anthropocene, to put it another way, should be grasped as a "totality" (as encompassing the "total" future history of the planet) but needs at the same time to be understood as an internally heterogeneous totality (as producing different effects when we approach it through the perspective of an individual human life within a particular historical order *and* from the perspective of humanity in its "species" being), then how can we understand those different orders to relate to one another? What form does their relation take when that relationship is simultaneously indexed to a singular situation (the situation confronting Collins Kusietey on the morning of March 7, 2010, in the village of Totope, Ghana) and to Chakrabarty's "planetary conjuncture"?

In addressing those questions, I return one last time to *Critique of Dialectical Reason* to discuss the ways in which a scaled-up version of Sartre's analysis of environmental change might cause us to regard climate change as a "counterfinality" (that is, as one of those apparently background environments of human activity that has been so transformed by prior human actions that it has trespassed the boundary between material background and human foreground to become a counteragent of history). I then turn to Timothy Morton's work to discuss the differences between this Sartrean (and still humanistically materialist) conception of counterfinality and Morton's new posthuman materialist reading of the Anthropocene as a "hyperobject." With that framework more clearly articulated, I return to Gilroy and Mbembe, to consider the ways in which their two versions of the black planetary map further develop the distinctions and potential points of convergence between these two materialisms, particularly as they both—similarly and quite dissimilarly—draw on the work of Frantz Fanon as pivotal to the questions of our planetary contemporaneity and future. In so doing, I ask how Quarmyne's arresting series of images—and the insight Sharpe's analysis provides in allowing us to regard those images as a form of "wakework"—can help make sense of the nature of the complex interrelationship between these two materialisms (Materialism I and Materialism II, as I have been calling them), subtending the dialectical interplay of the forces and forcings of our times.

In concluding, I return to the question of freedom that has been running through the entire text. In doing so I seek to respond to three key issues. The first is to directly address the question of why a project of freedom could or should continue to provide a central aspirational ground for a Black Atlantic Anthropocene political imaginary when quite arguably it is an *excess* of freedom (of capital growth, of consumption) that has helped create the catastrophic conditions of planetary climate change coming ashore around the world but with especially devastating effect in the outposts of the global postcolony. The second is to suggest that any answer to the question of freedom will necessarily vary depending on the scale at which the query is put. The challenge of freedom from exposure to the conditions of extreme vulnerability that climate change produces will, that is, look very different at the biographical scale of an individual human life (for which ideals of the dignity of personhood and an allied body of individual human rights remain crucial) than when it is posed at the zoëlogical scale of humanity's being as one planetary species among others (for which a more radically posthumanist conception of freedom more indebted to Latour's notion of the parliament of things or Mbembe's cosmological assemblages might provide a more significant response).

With that multiplicity of "freedoms" across the scales of the Anthropocene register in mind, I finally suggest that Quarmyne's work can help us frame a new conception of freedom (flowing from the subaltern particularity of the Black Atlantic but fully aspirational for a world, in its planetary totality, increasingly being remade in the image of the Anthropocene postcolony): a conception proceeding from the notion that freedom need not necessarily be held to begin and end with the freedom from constraint and the urge to self-protection—either individually or collectively (as it does for much Enlightenment thought)—but can instead be understood as emerging, simultaneously, from an ethic of being undone; an ethic of being decomposed and recomposed through our entangling set of relationships to the biological, and the zoëlogical, and the geological, and the cosmological orders and times of planetary life. Understood in this sense, to speak of freedom is to speak of an ethic reoriented *from* a solitary desire for immunity from the forces and forcings of our epoch *toward*, instead, a determination to refashion the biographical and nomological orders of our lives (our habits of dwelling, consuming, legislating) in relation to these forces and forcings of planetary life. From "freedom from" to "freedom toward": that, as I understand it, is one project for the future of our situation; one orientation for making something of what we have been made; one disposition toward the future that I believe we must pursue as we engage that dialectic of forces and forcings that has become one of the great hallmarks of our multilayered, multitemporal "epoch," planetwide, and as it comes ocean-crashing to ground on the vanishing edge of the Ghanaian shore.

2. History 4° Celsius

SEARCH FOR A METHOD

Scholars writing on the current climate change crisis are . . . saying something significantly different from what environmental historians have said so far. In unwittingly destroying the artificial but time-honored distinction between natural and human histories, climate scientists posit that the human being has become something much larger than the simple biological agent that he or she has always been. Humans now wield a geological force.
— DIPESH CHAKRABARTY

This is not only a historical age, but also a geological one.
— TIMOTHY MORTON

In "History and Dialectic," the concluding chapter of his 1962 volume *The Savage Mind*, Claude Lévi-Strauss famously outlines the operations of what he calls the "historian's code," the ordering principle by which the historical discipline seeks to bring human experience into a dialectical and potentially total relation with its fields of circumstance (or, as I will be stressing, its "situation"). The innocuous key to the historian's code, Lévi-Strauss argues, is chronology, and the fundamental materials of chronology are dates. However, he indicates, this apparently simple "chronological coding" "conceals a much more complex nature than one supposes when one thinks of historical dates as a simple linear series."[1] That is so for several reasons. Partially because not all dates are alike, either in their concentration of distribution along the "linear" axis of time or in the type of time they variously denote. In the first instance, "we use a large number of dates to code some periods of history; and fewer for others. . . . [T]here are 'hot' chronologies which are those periods where in the eyes of the historian numerous events appear as differential elements; others on the contrary, where for him [sic] . . . very little or nothing took place."[2] In the second instance, not only are dates thus unevenly distributed across "hot" and cold chronologies. Each date stands within its particular chronological domain not merely as an abstract, declarative, ordinal number, but as a performatively inflected "member of a class": "[thus] the date 1685 belongs to a class of which 1610, 1648 and 1715 are likewise members; but it means nothing in relation to the class composed of the dates: 1st, 2nd, 3rd, 4th millennium, nor does it mean anything in relation to the class of dates: January 23, August 17, September 30, etc."[3] That is so not simply because one class of dates denotes years, one millennia, and one days, but because each of these classes corresponds, in turn, to a "lower-" or "higher-powered" "scale" of history, with "biographical and anecdotal" history (measured, like diaries, in days) at the "bottom" of the scale and other "times" such as "the middle ages, antiquity, the present day" (measured variably in decades, centuries, and millennia) unevenly distributed across "different [scales of] power" above the diurnal zone of biographical time.[4] Consequently, what at first appears as a straightforward "general [historical] code" consisting in the sequencing of "dates" "ordered as a linear series," in fact articulates a highly complex procedure for managing, coordinating, and synthesizing a heteronymous array of "classes," domains, powers, and scales, "each furnishing an autonomous system of reference."[5] Lévi-Strauss notes that from this operation of the historian's code, as fully disciplinary as dialectical, the simultaneously

> discontinuous and classificatory nature of historical knowledge emerges clearly. It operates by means of a rectangular matrix:
>
> . . . .
> . . . .
> . . . .
> . . . .
> . . . .
>
> where each line represents classes of dates, which may be called hourly, daily, annual, secular, millennial for the purposes of schematization and which together make up a discontinuous set. In a system of this type, alleged historical continuity is secured only by dint of fraudulent outlines.[6]

Lévi-Strauss's fundamental purpose in mounting this critique was of course less a denunciation of historical method per se than a rejection of what he took to be the false equivalence Jean Paul Sartre had established between the "historian's code" and the "human order" in *Critique of Dialectical Reason*, particularly in the long introductory chapter (*Search for a Method*) in which Sartre had sketched his "progressive-regressive" method. In its elaborate reworking of Marx's famous dictum from *The Eighteenth Brumaire of Louis Bonaparte* ("men make their own history, but they do not make it just as they please in circumstances they choose for themselves, rather they make it in present circumstances, given and inherited"), *Search for a Method* had sought to effect a certain reconciliation between existentialism and Marxism by pledging fidelity to the proposition that the task of any properly dialectical philosophy of freedom (like any properly dialectical philosophy of history) lay in articulating the relationship between the domain of human action and the domain of historical necessity, the realms of making and of circumstance, of the actor and the "situation."[7] The error of Marx's inheritors, Sartre suggested, and the error existentialism was uniquely equipped to correct, was that of abandoning the original complexity of this "difficult synthesis" in favor of a reductive privileging of the objective over the subjective element within history, of the "situation" over man's capacity for "going beyond a situation"—thus abandoning the knowledge of "what [man] succeeds in making of what he has been made."[8] To correct the errors of this "superficial," "dishonest," and "lazy" Marxism, Sartre proposed his progressive-regressive method, whose "first moment" as David Sherman notes, "is really the regressive element, [which] works backward to analyze the particular historical factors that have gone into the construction of subjectivity, while the second moment, the progressive element, involves the way in which subjectivity synthesizes and transcends these factors in pursuit of its future projects."[9] Through the consequent "'internalization of the external' and the externalization of the internal . . . the subject, through its actions, freely makes the history that has made him [sic]" and so, as Sherman summarizes, allows Sartre to claim that he has found a way to "discard neither freedom nor history."[10]

For Lévi-Strauss, the problem with this lay not in Sartre's turn to Marx but in the internal flaws of the progressive-regressive method and, more vitally, in the false and total equivalence it established between the human order and the historian's code. Far from resolving the relation between the human and the situation, freedom and necessity, Sartre's method, according to Lévi-Strauss, had fundamentally misconstrued the nature of humanity's dialectical entanglements. Minimally, this flowed from Sartre's misunderstanding of the true nature of historical time, which, Lévi-Strauss indicates,

Sartre continued to treat as a linear arrangement of equivalent dates rather than as that matrix of scales of "power" (each with its own "autonomous system of reference") he had outlined. Moving ever further "backward to analyze the particular historical factors that have gone into the construction" of a singular subjective position, Sartre's method promised to place the human actor first in the day of his or her making, then in the year, then in the decade, then in the century—until the operation had been completed, and an entire historical "situation" had been established as that zone of necessity from which a human actor, having been made, might then find the conditions for a progressive (re)making of what has been made.

To promise such a thing, Lévi-Strauss argues, is simply to obscure that biography is neither periodicity nor epochality, that coming to the end of the line of any single subject-situating "class" of dates, the regressive method finds itself obliged to leap levels from one domain of power to another, while nevertheless obscuring the fact that any leap has been made or that in making that leap it has found itself obliged to mix or elide the distinctive modes of intelligibility giving each scale of history its marked character. The biographical subject of days and years, to put things another way, is not simply the subject of a period-milieu measured in decades, then the subject of a modernity, or a Renaissance, or an antiquity measured in centuries, and finally the subject of an enframing epoch of human "historical time" measured in millennia and set off (from the dawn of the invention of agriculture) from "pre-history."[11] Biography, period, era, and epoch (to name but four of the historical code's "classes" of power) may be (indeed must be) related, but they are not the same, and the progressive-regressive method entirely fails to account for their difference.[12]

This, though, does not yet touch Lévi-Strauss's central argument and critique. For even if Sartre had produced an account of the diversified logics and orders of power of the multiple scales of historical time *and* a theory of their relation to one another, he still would have erred, the anthropologist suggests, by assuming that the problem of the human is exhausted by an examination of the relation of singular existence[s] to the grid of intelligibility provided by the historian's code. I say "exhausted" advisedly. To reiterate, Lévi-Strauss's critique is not a critique of history *tout court*. It is not a critique of Sartre's desire to put the human in a dialectical relationship with material structures, processes, or events. (Lévi-Strauss accepts the need to do so, provided one can simultaneously mark the internal heterogeneity of history's scales of power, intelligibility, and order.) Rather, he insisted, Sartre's mistake was to treat the historical matrix as if it was the *unique* and *total* matrix of human "circumstance" and, therefore, to reduce the challenge of dialectical reason's accounting of human being to a matter of emplotting the "human" as one or other solitary point within the grid of history.

In contrast to that history-exclusive model, Lévi-Strauss maintains, a fully dialectical account of humanity must both attend to "history" *and* address what the historical code fails to hold within its grid. And to do this, he insists, philosophy must find an epistemology for simultaneously working through the historical and for "getting outside history"[13]—not once, but twice.

First, he says, we must find a route for exiting history "to the bottom . . . that is to say to an infra-historical domain in the realms of psychology and physiology."[14] And secondly, we must get outside history to "the top. . . . into the general evolution of organized beings, which is itself explicable only in terms of biology, geology and finally cosmology."[15] As I parse it, Lévi-Strauss in this way proposes adding to the historian's code an infra-historical and a supra-historical domain that collectively (and in their mutual exchanges with the historical domain) realize a dialectic of human existence that could be schematized thus:

Infra-historical Domain

Psychology
Physiology

Historical Domain

Supra-Historical Domain

Biology
Geology
Cosmology

It is only by accounting, simultaneously and continuously, for the relation between the scales of circumstance proper to "historical" time, and the infra-and supra-historical domains of psychology, physiology, biology, geology, and cosmology that one can truly provide an adequate account of the human dialectic of freedom and necessity, of the actor and the situation, of that full range of "properties" from which the human situation is composed.[16] To accomplish this, Lévi-Strauss thus argues, one must add to the historian's code a heteronymous order of knowledge: an order of thinking and knowing capable of blending psychology, cosmology, biology, physiology, and geology; that order which Lévi-Strauss identifies as *savage*. And for that, he insists, philosophy requires, as a supplement to the science of history, the science of anthropology, a science not only descriptive but imitative of the *savage mind*'s "intransigent refusal . . . to allow anything human (or even living) to remain alien to it," a science that discovers in that openness, in that willingness to "undertake the resolution of the human into the nonhuman," in that capacity to effect "the reintegration of culture in nature and finally of life within the whole of its physico-chemical conditions," the "real principle of dialectical reason."[17]

I have returned to "History and Dialectic" for several reasons. If the first, and most obvious, is that almost exactly fifty years after Levi-Strauss set the terms of his debate with Sartre in this way, he seems to be on the point of a conclusive victory, then I do not mean by that that we are on the verge of a return to structural anthropology (though there are, certainly, signs of the emergence of a range of structural biologies across numerous spaces of contemporary intellectual life); or that we have been living unaware, in all the years since 1962, in a long Sartrean moment that is finally coming to an end (though, as James Chandler has noted, in the line from Marx, to Sartre, to Fredric Jameson, there has been perhaps a greater Sartrean influence on post-existentialist, materialist epistemologies than we sometimes recognize).[18] Rather, I wish to call attention to the uncanny precision with which, in staging his quarrel with Sartre, Lévi-Strauss anticipated not only the emergence of the "geological" as a figure of contemporary critical theory but also a far broader major current of critical discourse predicated on the imperative of getting thought materially "outside history." And while it is absolutely the case that that contemporary discourse has been shaped by the transition from structuralism to poststructuralism in the intervening years, it is nevertheless remarkable (perhaps in a way analogous to the remarkable persistence of categories of Saussurian structural linguistics into the problematics of deconstruction) that this contemporary turn outside history has proceeded in exactly the two directions Lévi-Strauss counseled: "beneath" history and "above" it.

On the materialist turn "beneath" history—the turn toward what Lévi-Strauss identified as the conjoined infra-historical domain of "psychology and physiology"—I have in mind, to take just one example, the explosive recent turn toward the neuronal. The proliferation of neuro-politics, neuro-economics, neuro-philosophy, and neuro-humanities culminates a turn toward the borderlands of brain and mind, of affects and emotions, of synapses and desires; a turn toward that zone of "plasticity," which in its capacity, as Catherine Malabou observes, to "give shape to," "take shape from," and "explode the shape of" human consciousness has come to demarcate a newly visible psycho-physiognomic terrain of the modern dialectic of freedom and necessity; a turn which, in that sense (and this is a point to which I will return), does not so much obviate Sartre's motivating Marxian problematic of the subject and the situation, of making and being made, as extend that problematic in materially new and previously unrecognized directions.[19] To revise Marx's dictum: here the brain *is* history; or, as Daniel Lord Smail has it, the evolutionary history of the brain is the deep history that is our circumstance, the history we must confront if, in Malabou's suggestive formulation, we wish to "do" something with our brain: if we wish to make something of what it has made us.

On the corresponding and simultaneous material turn "above" history—the contemporary supra-historical turn toward Lévi-Strauss's "general evolution of beings . . . explicable only in terms of biology, geology, and cosmology"—the field of debate is as crowded and brightly lit as the skies of the cosmos. To list but a very few of the more prominent names: Donna Haraway, Jane Bennett, Bruno Latour, Tim Morton, and Quentin Meillassoux have all, in overlapping and discrete ways, sketched a series of companionate, vibrant, thing-political, enmeshed, and ancestral zones of strangely strange, biotic, nonhuman, geological, and cosmological "actants" without whose consideration any future raising of the question of the human and its fields of circumstance (its "situation") will prove inadequate.[20] The constellation of projects these scholars' works have outlined, to repeat, does not converge on a single new line of materio-epistemological insight. Meillasoux in particular, together with the more general speculative-realist, non-correlationist, object-oriented philosophy he alternately articulates and is made to stand in for, certainly occupies a registrably disjunct place relative to the other scholars' pursuit of a post-Kantian but still co-relational account of the "human" as one form of speciated-being, among others, within the evolutionary, ecological, geological, and cosmological order of things.

Despite these differences, we can discern within this bundle of critical thought (both in its infra-historical and supra-historical domains) the collective working out of something like what Foucault called a *dispositif*: the coming into operation of "a thoroughly heterogeneous ensemble" nevertheless characterized by "a system of relations" giving rise to an internally riven and singular "formation which has as its major function at a given historical moment that of responding to an *urgent need*."[21] That need, in Latour's terms, is the need of finally having done with the "bicameral parliament" of modern thought; the need to move beyond the false conception that there is a domain of nature on the one hand and a domain of politics (and culture, and history) on the other; the need to frame an understanding of the non-oppositionality of nature and culture; to articulate what Haraway calls a theory of nature-cultures; or, indeed, as Lévi-Strauss explains in "History and Dialectic" (significantly turning from his earlier work), to finally "undertake the resolution of the human into the nonhuman" and "the reintegration of culture in nature"—and, thereby, to supersede the critical and philosophical dominance of the historian's code as sole arbiter of the human condition.[22]

One mildly paradoxical reason to return to the Sartre/Levi-Strauss debate is, therefore, to discern the ways in which it can be seen to throw into visibility an at least five-decade-old period of critical time (spanning the transition from late-stage structuralism to late-stage post-structuralism) predicated on the shared urge of getting thought *outside history*. That spotlight (to begin to close the circle to my point of beginning) allows us to observe the ways in which a discourse on the Anthropocene (dateable, in one sense, from Crutzen's initial published use of the term in the 2000 International Geosphere-Biosphere newsletter) belongs, coincidentally, to that more extensive, if increasingly "hot," chronology of critical time running, minimally, from the 1960s to the present. Another and more fully paradoxical reason for returning to Sartre and Lévi-Strauss is that—even as the discourses on and of the Anthropocene, the neurological, and the ecological finally begin to shift us into a time after the dominance of the "historian's code"—that shift, as I have already intimated, does not so much dissolve the problem Sartre's neo-Marxian project was designed to address as massively expand it.

If Sartre's dilemma, in other words, was to find a mid-twentieth-century existentialist response to the nineteenth-century Marxist question of whether a human project of freedom could survive the "difficult synthesis" of the subject and the situation, then this extra-historical turn (across its multiple domains) has again found itself obliged to pose the question of the human, and of freedom, though now at a yet higher order of philosophical complexity. It has found itself again wrestling with the question of what shape freedom (Mallabou's "alter-global," or Morton and Latour's "democracy" and fashioning of the future) might take when we consider the problem of future-fashioning as arising not only from the foundational Marxian dialectic of the subject and the (historical) situation but from within a second dialectic of the "situation" as, itself, multidimensionally infra-historical, historical, and supra-historical.[23]

Supplementing the cultural, economic, sociological, and political conception of circumstance that the historian's code makes available, this second-order dialectic finds the situation of the human both collapsing inward and exploding outward, veering simultaneously synaptic and planetary. Ramified into the force field of those spaces, the question of how to go "beyond" the "situation" (the question of how to "succeed in making [something] *of* what has been made") does not, thus, so much leave behind the Sartrean question (or, really, the Marxian/historical materialist question) as multiply and disperse its urgency across the neurological, nomological, geological, and cosmological fields. Recurring, multiplying, expanding in this way, that question thereby highlights the need not merely for an answer (or a set of answers) but for a method of coming to answer, a method for thinking the relation of the "human order" to all these domains (individually and in common). The compound of discourses forming the extra-historical *dispositif* I have limned in its five-decade leap from *The Savage Mind* to the present implies the need for a new search for a method: one that will take as its starting point an investigation of the multi-scaled, simultaneously historical, infra-historical, and supra-historical "situation" in which we find ourselves—and from which we might take the cues toward the task of our own project of freedom and democracy.

It is on this exact point that Chakrabarty's "The Climate of History" bears returning to, precisely because, like Sartre's *Search for a Method* and Marx's *The Eighteenth Brumaire of Napoleon Bonaparte*, it directly foregrounds the dialectic of freedom and the situation—and because in taking up the crisis of freedom in the age of the Anthropocene, his theses begin to frame the outline of a new concept of history for this expanded post-humanist historical situation and also, provisionally, to provide a mode of responding (through the "negative universal" of our species-being) to what that situation is making of us.[24]

The first step toward this new mode of understanding, Chakrabarty argues (in similar spirit, if different effect, to the line of thought stretching from "History and Dialectic" to *The Companion Species Manifesto*, *The Politics of Nature*, *Vibrant Matter*, *The Ecological Thought*, and *What Should We Do with Our Brain?*), is to collapse the age-old humanist distinction between natural history and human history. He takes as a starting postulate that we have entered an era in which human and natural forces have merged, and human action has taken on the qualities of a "force of nature" in shaping the long-term geological future of the planet.[25] Chakrabarty's immediate corollary to that point is that postcolonial studies, therefore, requires a fundamental rethinking of many of the key values underlying its prior methodological commitments, key among them its anti-universalism, inadequate now, he suggests, to a universal challenge of planetary existence, and so at least provisionally worth replacing with a new "negative universal," flashing up in that moment of danger, which is climate change: the new universal of "species" being.

This is well known and already widely commented on in a range of critical responses to Chakrabarty's arguments.[26] What might be less recognized is that even as this new historical thought of the Anthropocene requires setting on hold long-standing critiques of the universal, its full methodological implications also seem to demand leaving behind, as an outmoded relic of self-enclosed "human history," an equally foundational investment in and critique of post-Enlightenment projects of human freedom running through much postcolonial and allied bodies of critical theory. At stake is the counterplay of Enlightenment and subaltern conceptions of freedom, justice, and democracy that lay at the core of Chakrabarty's celebrated earlier study of the complex inter-animation of what he previously called History 1 and History 2.

By History 1, as he details in *Provincializing Europe: Postcolonial Thought and Historical Difference*, we should have in mind an Enlightenment-inspired, progressive theory of history (classically associated with a post-eighteenth-century "historicism") and an attendant politics of rights-based citizenship and democracy, a theory and a politics that postcolonial theory cannot simply reject but must, instead, mark as simultaneously "indispensable and inadequate."[27] "Indispensable," not only because the project of securing full and equal rights of participatory citizenship and the protection of the individual against the power of the state has been, and must remain, one of the key elements of any vibrant anti-imperial and postcolonial politics, but also because the analytic procedures through which Marx derived those categories of his thought, such as "abstract labor," from which he was able to derive not only a descriptive account but a critique of capital, depend on the universalizing conceptual legacy of History 1. If we are to have universal critiques of capital and its role in the modern projects of empire—which we must, Chakrabarty argues, if for no other reason than that "grasping the category 'capital' entails grasping its universal constitution"—then on this ground also we cannot do without History 1.[28]

Thus indispensable, History 1 is also "inadequate," for at least three reasons. First, because the teleological code underpinning History 1's "historicism" has repeatedly posited "historical time as a measure of the cultural distance . . . assumed to exist between the West and the non-West" and in doing so has furnished endless alibis for the imperial civilizing mission.[29] Second, because while History 1 equates modernity with a narrative of unilinear global progress *and* with an abstract, analytic, and entirely secular epistemology, subaltern politics (even as it assumes the indispensability of rights-based democratic norms) "has no necessary secularism about it." "It refuses to "take the idea [of a] single, homogeneous, and secular historical time for granted," and "continually brings gods and spirits into the domain of the political."[30] And third, because in addition to the "analytic" critique enabled by the universal protocols of History 1, postcolonial politics thus requires a "hermeneutic" critique arising from these subaltern "life-forms" and their insistent acts of interrupting the secular politics of History 1, and interweaving with its Enlightenment ideals alternate "imaginations of socially just futures for [the] human."

In addition to History 1, Chakrabarty concludes, we therefore require a further concept of history: one that recognizes that "historical time . . . is out of joint with itself"; that the human is not "ontologically singular"; that "gods and spirits [are] existentially coeval with the human"; that in addition to an abstraction-driven critique of capital (and empire) we require this coincident, affectively rich, and anthropologically differentiated order of critique.[31] His name for that concept is History 2, a form of history that, crucially, is not the binary "other" of History 1, or its archaic antecedent, but is "better thought of as a category charged with the function of constantly interrupting the totalizing thrusts of History 1."[32] In their relation with one another, History 1 and History 2 do not, therefore, express an antinomy. Rather, Chakrabarty argues, they reveal that the time of democracy (the time of the struggle for "socially just futures"), far from expressing a single, universal, and unidirectional chronology, exists as a set of "time-knots" in which we live. These knots braid together the secular and the nonsecular, the universal and the particular, the analytic and the hermeneutic, Enlightenment and subalternality in intertwining and separating projects of emancipation and justice.

What becomes of this complex interplay of History 1 and History 2 in the turn to the crisis of climate? What becomes of their undecidable complementarity in what we might now call *History 3*'s new theoretical accounting of the advent of the Anthropocene?

At first glance, History 1 and History 2 seem to survive the transition to this urgent new historiographic regime, particularly as Chakrabarty's new historical mode acknowledges the continuing relevance of attempts to address the role of capital in precipitating the catastrophic rise of carbon emissions. As he indicates in his third thesis: "Analytic frameworks engaging questions of freedom by way of critiques of capitalist globalization have *not* in any way become obsolete in the age of climate change. If anything, as [Mike] Davis shows, climate change may well end up accentuating all the inequities of the capital world order if the interests of the poor and vulnerable are neglected."[33] Between the continued relevance of a global/universal critique of capital and the need to account for the differential exposure of the vulnerable and poor to the devastating effects of climate change, there seems to be room within History 3 for both Histories 1 and 2.

But that is not what Chakrabarty seems finally to assert. History 1 and History 2 may not be "obsolete," but they are also not fully incorporated within his new line of thought. Like the "theories of globalization, Marxist analysis of capital, subaltern studies, and postcolonial criticism" that, he indicates, have ultimately proven inadequate to "making sense of this planetary conjuncture within which humanity finds itself today," they are instead more hauntologically prior to this new approach, reappearing in his argument as the residue (or, in an ironic twist of Robert Young's terms, as the "remains") of a "[still]-present historiography of globalization" that gives every appearance of being on the way to becoming obsolete as the walls of "human history" are "breached." This historiography exists in a condition of fundamental "difference" with (and quasi-historicist anteriority to) "the historiography demanded by anthropogenic theories of climate change."[34] The grounds of that difference are, for Chakrabarty, multiple. Key among them, however, is that while both History 1 and History 2 can be experienced, the new geophysical form of "human collectivity" brought about by the Anthropocene escapes our capacity to "experience."[35] By which, as he makes clear, it escapes "our" capacity as humans to experience what it means to breach the boundaries of human ontology, to traffic with (and as) the nonhuman, to have become humanly nonhuman and nonhumanly human.

This is a remarkably complex statement, all the more remarkable because, in advancing it, Chakrabarty seems not only to be moving from the older Marxist materialism with which his work has long been in conversation (what I am calling Materialism I) to the "new materialism" (Materialism II) articulated by Bennett and others, but because, in doing so, he seems to elide some of the vital insights of his prior analysis of the interanimation of History 1 and History 2—particularly the insight that human history has never been ontologically singular, that it has always involved the traffic between the human and the non-human. Perhaps another way of phrasing things might have been that the crisis of the Anthropocene does not so much demand that we brace ourselves for looking beyond the undecidable interplay of History 1 and History 2, but instead that we expand our sense of the ontological plurality of the human; that to those supernatural actors and agents with whom we have earlier seen the human to be coeval, we must now also recognize the post-natural actors, agents, and actants of cyclones, heat waves, and melting ice; that perhaps we do not need so much a History 3 as an expanded History 2, more extensively interrupting and modifying History 1.

Yet that is not the direction Chakrabarty's argument takes. Rather, as he turns his attention to the challenges that the transformation of human being into species being puts to the question of freedom, he suggests that the time has come for a fundamental reconsideration of the linkage between philosophical critique and the grand modern project of freedom. And, entirely consistently, he does so not because he wishes to return to or intensify his earlier investigation of the mutually modifying exchanges of Enlightenment and subaltern conceptions of freedom but for a devastatingly empirical set of reasons. Because, as he puts it: "The Mansion of modern freedom stands on an ever-expanding base of fossil fuel use. [Because] most of our freedoms so far have been energy-intensive."[36] Because "whatever our socioeconomic and technological choices, whatever the rights we wish to celebrate as our freedom, we cannot afford to destabilize conditions (such as the temperature zone in which the planet exists) that work like boundary parameters of human existence."[37]

Despite their parsimony, the consequences of these statements are far-reaching. At the very least, they lead to a series of assumption-troubling questions. "[H]as the period from 1750 to now been one of freedom or that of the Anthropocene?" "Is the Anthropocene a critique of the narratives of freedom? Is the geological agency of humans the price we pay for the pursuit of freedom?" To all of which Chakrabarty responds: "In some ways, yes. As Edward O. Wilson said in his *The Future of Life*: 'Humanity has so far played the role of planetary killer, concerned only with its own short-term survival. We have cut much of the heart out of biodiversity.'"[38] Central to all these questions, and to Chakrabarty's response, is the mournful but resolute understanding that while History 1 and History 2's projects of freedom (Enlightenment, subaltern, or some co-modifying hybrid of the two) may have posited freedom as the end point of history, such freedom has instead proven to be the portal to something else, something catastrophic. Indeed, the tragic secret knowledge of what I am calling Chakrabarty's History 3 (the reason this historiographic mode may feel obliged to keep History 1 and History 2 outside itself) may well be that modernity and postmodernity's great projects of freedom (Enlightenment and counter-Enlightenment) *are* the catastrophes leading into one of Agamben's periods of "ultra-history," the catastrophe leading to another end of history, to an image of the end metonymically figured not only by the image of a single vanishing species, but by virtually all the tipping-point, threshold-crossing, cascading images of the 4°C world: the image of death, the image of extinction.

This, finally, is the key code of this new historical mode: that even as it foregrounds the question of freedom, it does so no longer in order to orient us toward a future measured against the promise of freedom but, instead, to direct us to (and desperately against) a future marked by the threat of extinction. In doing so, it no longer tries to derive "socially just visions of the future" from the promise of coming democracy but from a collective, planetary being-toward-death. That does not mean giving up on justice. But it does mean that for this manner of conceiving history, justice must be construed in an entirely different mode.

Whereas for the enlightened and subaltern political projects of History 1 and History 2 (projects for which the subject appears under the alternating/overlapping guise of the "citizen" and the "peasant"), justice comes to occupy the contested and undecidable zone between what the law mandates and what "forms-of-life" create; for a historical method oriented toward extinction and emerging under the sign of a subject appearing in the guise of "species," justice leaves behind law and anthropology and now occupies itself with questions of ontological transformation and survival—questions of how justly to survive a transformed ontology of being.

From freedom to extinction; from the citizen and the peasant to the species; from law and forms-of-life to ontology—these, I am suggesting, are the deep code of the transition from History 1 and History 2 to History 3 emerging in Chakrabarty's work.

Let me be clear. I believe there is significant value in this transition, in the attempt to imagine an ontological politics for the deep future of the planet under the sign of species.[39] Nevertheless, it is worth—and necessary—asking whether the power of that insight is such that this is the sole choice now available to us?

Or, reformulated: Is an orientation to extinction unitary or does it, instead, contain within itself other orientations? If not oriented exclusively toward freedom as it has been construed in major currents of occidental political theory from the Enlightenment onward, then perhaps, as Chakrabarty's mildly open formulation glancingly hints, toward some other ways of conceiving freedom? Are there other orientations toward the future that might continue to be routed through the cultural and historical, the social and aesthetic archives of the postcolony—while also being routed, as Latour and Morton in their different fashions have it, through a radicalized concept of democracy, through the mesh of "strange-strangers" and the "parliament of things"? Or, as Jane Bennett has expressed it, through a renovated conception of the Spinozan "conatus," through a reimagined "endeavor to persist in being"—now not through the will to security but through "the will to belong," "as one species on the planet among numerous others"?[40]

If there is space for such multiplicity within the political being of the Anthropocene, space for more than the "negative" universality of humanity's species being, space and room for the possibility that extinction is not the sole copula between the "subject" and this "situation," room, as Bennett would say, for alternate "projections of a fittingness between humanity and the future," then, to take another vital point from Chakrabarty's work, that will be because our conception of politics is bound up with our conception of the nature of historical time, and because, recognizing this, we have warrant to extend his foundational insight that neither human ontology nor the ontology of time is singular but plural, warrant to maintain that time (including the time of the Anthropocene) continues to be out of joint, that the time of time-knots is not over.[41]

Or—to move backward from Chakrabarty to the text by Lévi-Strauss with which I began this section—let me put it this way. If there is reason to believe that both the history and the politics of the Anthropocene are not unidirectionally oriented but multidirectional, that will be because, once again, the historian's code remains inadequate to an accounting of our "situation." The Anthropocene is more than a name for a new chronology or a new set of historical dates, now measured longhand in millennia rather than in mere decades or centuries but, still, like the chronologies of old, moving inevitably, teleologically, in a single direction. This new supra-, ultra-, or extra-historical moment we inhabit is one composed of multiple scales, orders, and classes of time (historical and extra-historical; subaltern, hermeneutic, and ontic) and multiple corresponding orientations to the possibility of the (just) future fashioning of those times. To revise my earlier suggestion: perhaps it is not a matter of returning, in expanded form, to the weave of History 1 and History 2, but of slip-knotting them together, with History 3, into one more braided order of time. This fourth order of history, measured both in dates and in degrees, in times and temperatures, might prove adequate to the temporal and cultural and ontological multiplicity of the 4°C world, serving as a historical, infra-historical, and suprahistorical order of Enlightenment, subaltern, and geophysical time that I am calling History 4°.

Let me recapitulate. I have suggested that the 4°C world is temporally, culturally, and ontologically multiple and that to its multiple temporalities and heterogeneous ontology there is a correspondingly multiple and heterogeneous set of orientations to the future. I have further suggested that there are two key (and still unanswered) questions that follow from those propositions. What are the strands of time of which that world is composed? And what is the method for unbraiding those strands, for unraveling the plural situation of what I am calling History 4°?

We need now to respond to those questions, and in responding to them we must multiply them just a little. In place of the two questions, let me then propose four queries that stand as complements, of a sort, to the four theses that Chakrabarty outlines in "The Climate of History."

- What are the orders of time of the "situation" of History 4°?
- How do those temporal orders relate to one another, and where might we look for models of their relationship?
- What projects of future-fashioning do they variously imply?
- How, as they orient us to the future, do they reopen or renovate the question of that dialectic of forces and forcings—of human and extra-human history—that I argue is the hallmark of our profoundly mixed "epoch" of time?

With that full bundle of questions in mind, let me return again to Sartre, not to *Search for a Method* but to *The Family Idiot*, or, more precisely, to the preface to that text, a brief two-page statement that, as it turns out, is not so much the preface to a new work as a bridge between the two. The statement is worth citing almost in full:

> *The* Family Idiot *is the sequel to* Search for a Method. *Its subject: what, at this point in time, can we know about a man? It seemed to me that this question could only be answered by studying a specific case. What do we know, for example, about Gustave Flaubert? Such knowledge would amount to summing up all the data on him at our disposal. We have no assurance at the outset that such a summation is possible and that the truth of a person is not multiple. The fragments of information we have are very different in kind: Flaubert was born in December 1821, in Rouen—that is one kind of information; he writes, much later, to his mistress: "Art terrifies me"—that is another. The first is an objective, social fact, confirmed by official documents; the second, objective too, when one sets some store by what is said, refers in its meaning to a feeling that issues from experience, and we can draw no conclusions about the sense and import of this feeling, until we have first established whether Gustave is sincere in general, and in this instance in particular. Do we not then risk ending up with layers of heterogeneous and irreducible meanings? This book attempts to prove*

that irreducibility is only apparent and that each piece of data set in its place becomes a portion of the whole, which is constantly being created, and by the same token reveals its profound homogeneity with all the other parts that make up the whole. *For a man is never an individual; it would be more fitting to call him a universal singular. Summed up and for this reason, universalized by his epoch, he, in turn, resumes it by reproducing himself in it as a singularity. Universal by the singular universality of human history, singular by the universalizing singularity of his projects; he requires simultaneous examination from both ends. We must find an appropriate method.* I set out the principles of this method in 1958 and will not repeat what I said; I prefer to demonstrate whenever necessary how this method is created through the very work itself in obedience to the requirements of its object. . . . Now we must begin. How, and by what means? It doesn't much matter: a corpse is open to all comers. The essential thing is to set out with a problem.[42]

While it has been fundamental to my argument that the "situation" of History 4° is precisely one in which we can no longer answer the question "what, at this point in time, can we know about a man?" (or let us say, instead, a "person") either by referring that question solely to the "singular universality of human history" or by solely leaving "human history" behind, then it has also been my argument that as the question of human knowability now finds itself equally referred to as the "universalities" of biological, geological, and cosmological history, the search for "an appropriate method" of resolving (like Sartre, but for our "extra-historical" moment) the Marxian dilemma of what it means to grasp and make history under circumstances, not of our choosing, does not end but explodes in significance and complexity. Reading the Sartrean search for a method against the particular method Sartre details—or, alternately put, reading Sartre, as I have attempted, through Levi-Strauss and (as I have begun to detail and will more fully outline below) through that extra-historical line of thought stretching from Levi-Strauss's "History and Dialectic" to Latour's *Politics of Nature*, Morton's *The Ecological Thought*, Bennett's *Vibrant Matter*, and Chakrabarty's "The Climate of History"—thus presents itself as one potential key to answering, with Sartre and against him, this question:

What, at this point in time, can we know about a person and, by virtue of knowing that person, know of the Anthropocene "epoch," the Anthropocene situation, she or he simultaneously "sums up" and summons toward a future?

Ultimately this allows us a shift to the question we have had from the beginning:

"What, now, can we know of Collins Kusietey? What now can we know of this young boy, standing on that Ghanaian shore, and of the arriving and coming epoch he represents and portends?"

But we are not yet ready to respond to that question. There is, first, further work to be done in clarifying the method of responding to it. Minimally, we need to address the key propositions behind it: the key philosophical postulates at the heart of Sartre's framing argument—one regarding the nature of the person, the other regarding the relation of the person to the situation.

What is Sartre's first postulate? In brief, that a human life, in whatever "epoch" it finds itself, is not composed of "layers of heterogeneous and irreducible meanings," but is, instead, totalizable; that "irreducibility is only apparent . . . each piece of data set in its place becomes a portion of the whole, which is constantly being created, and by the same token reveals its profound homogeneity with all the other parts that make up the whole."[43] At first glance, this does not at all conform with the argument I have been advancing, particularly as it seems to rely on Sartre's outright rejection of the notion that "the truth" either of a "person" or a "situation" is "multiple." Vitally, however, what Sartre rejects is not multiplicity per se but a particular mode of multiplicity, a version of multiplicity in which the "heterogeneous layers" of a life are radically autonomous from and opaque to one another, radically irreducible to some common form of articulation. And this, also, is not what I understand multiplicity to entail. Rather, as I have hoped to register in the second of my framing queries, the question of the multiple is not the question of irreducibility, but of the relation of times, scales, and ontologies to one another within the co-relational human and nonhuman totality of what I am calling History 4°. While the form of that relation might not be that of a "universal singular" as Sartre conceives of it, the conviction that there is a relational totality of the 4°C world, a relational totality of this time (however internally multiple and heterogeneous it may be), is fundamental to the method I am seeking to explore.

What form, then, does that totality take? For Sartre, in the terms of his second organizing postulate, it is the dialectical form of the "universal singular," the form humanity takes as at once something constituted by its moment and something constitutive of that moment; the form of the human as something fashioned, made, summed up by its "epoch," and the equal, countervailing form of the human constituting itself as an epoch-fashioning force through the tasks, "projects," and orientations toward the future it takes on. It is on this exact point that the method of thinking the relation of the "person" to the "situation" that I have in mind (the method of thinking and responding to History 4°'s dialectic of forces and forcings) both extends Sartre's neo-Marxian, quasi-Lukácsian insights and departs radically from them. That is because, while for Sartre, as he quite clearly states, the epoch, however vast, is always finally and only the epoch of "human history," the epoch I have in mind (the Anthropocene epoch within which humanity continues to discover itself as something constituted and something constituting) is not the singular epoch of human history, but a mixed epoch—an epoch of the human and the nonhuman; an epoch of the historical, the infra-historical, and the supra-historical.

Of what, then, does that mixed epoch consist? What constitutes its heterogeneous and related layers? What form does their relation take?

More immediately: how can we go about answering those questions?

Sartre answers them by turning to a particular case, a particular work, a problem within that work, and a question that problem raises.

Is that the way to proceed?

Perhaps, although there is nothing exclusively Sartrean about this. Much twentieth-century Marxian philosophical and aesthetic method proceeds in exactly this fashion. Take, for instance, the other major figure inspiring not only the methodological protocols of Chakrabarty's "theses" but my own thinking on that mode of counter-historicist thought adequate to the situation of our time, Walter Benjamin, whose seventeenth thesis on the "Philosophy of History" includes these justifiably famous observations:

> The materialist writing of history for its part is based on a constructive principle. Thinking involves not only the movement of thoughts but also their zero-hour [Stillstellung]. Where thinking suddenly halts in a constellation overflowing with tensions, there it yields a shock to the same, through which it crystallizes as a monad. . . . In this structure [the historical materialist] cognizes the sign of a messianic zero-hour [Stillstellung] of events. . . . He perceives it, in order to explode a specific epoch out of the homogenous course of history; thus exploding a specific life out of the epoch, or a specific work out of the lifework. The net gain of this procedure consists of this: that the lifework is

preserved and sublated in the work, the epoch in the lifework, and the entire course of history in the epoch.[44]

At first glance, this looks remarkably like Sartre (or more accurately, Sartre, two decades later, looks remarkably like this), and there are meaningful similarities. For Sartre, the sequence runs from the question to the problem, from the problem to the work, from the work to the totality. For Benjamin, the method flows from the "zero-hour" to the work, from the work to the lifework, from the lifework to the epoch, and from the epoch to the entire course of history. In either case, the movement from question/zero-hour to totality/entire course of history proceeds step by step through a moment of interrogated or arrested thought, to the interrogating/arresting work, to the total historical ensemble that work sums up or constellates. But there is a crucial difference. For Benjamin, that last step, that final turn toward "the entire course of history," does not limit itself to human history—either in what that entirety encompasses or in the mode of perceiving it.

As he makes evident in his immediately ensuing (if little discussed) eighteenth (and final) thesis, the ultimate horizon of historical thought must contain not only the history of "civilized humanity," but the longer history of "homo sapiens," the still vaster history of "organic life on earth," and then, beyond that, the truly deep cosmological history of the "universe." The full thesis, one of Benjamin's very last works, written in Vichy Paris shortly before his failed flight to Spain and the United States in the late summer of 1940, reads so:

> "In relation to the history of organic life on Earth," notes a recent biologist, "the miserable fifty millennia of homo sapiens represents something like the last two seconds of a twenty-four hour day. The entire history of civilized humanity would, on this scale, take up only one fifth of the last second of the last hour." The here-and-now, which as the model of messianic time summarizes the entire history of humanity into a monstrous abbreviation, coincides to a hair with the figure, which the history of humanity makes in the universe.[45]

Massively expanding the scale of the "here-and-now" to encompass the several thousand years of civilized humanity's recorded history (what I am inclined to call its "nomological time"), and then reminding us that even that vastly extended time frame of the historical present accounts for only a fraction of the history of humanity's species life (what we may call its "biological time"), which accounts only for a tiny margin of "organic" life (zoëlogical time), and that that organic life, in its turn, barely registers on the clock of the time of the universe (cosmological time), Benjamin thus completes his "Theses" and sets a final critical-methodological framework around the course of his own biographical time by extending the philosophy of history into those exact domains of the infra- and supra-historical that Lévi-Strauss subsequently introduced to his dialectic of history. And then, in a fashion entirely consistent with the course of his "lifework," Benjamin also includes one more temporal mode, a last order of time to complete the set of inward and outward unfolding temporal scales I understand to constitute the now-being of History 4°.

For if the eighteenth thesis's outermost sphere of historical thought at first glance appears (as in Lévi-Strauss's "History and Dialectic") to be the cosmological, then, Benjamin indicates, the cosmological (which encompasses the geological, which encompasses the zoëlogical, which encompasses the biological, which encompasses the nomological, which encompasses the biographical) is itself (are all themselves) rendered visible in relation to yet one more margin of time: the margin of messianic time, the theological margin of humanity's traffic with the nonhuman. To move from the zero-hour of a life, of a case, of a work, toward totality is thus not to move only from a "person" to the "universal singularity of human history" (to move, restrictedly, from the biographical to the nomological within an exclusive framework of "human history"), but to see how those "chronologies" (of a life, of a nomothetically ordering political order) are encompassed within at least five additional, overlapping, and related layers or scales of "nonhuman" time: biological time, zoëlogical time, geological time, cosmological time, theological/messianic time.

With that lesson from the zero-hour of an exemplary life and lifework in mind, we are now one step closer to re-posing, in more truly Benjaminian than Sartrean mode, a question from our own quite different "zero-hour," our "moment of danger" according to Chakrabarty.

What, at this point in time, can we know, neither of Gustave Flaubert nor of Walter Benjamin, but of Collins Kusietey?

But we are still not quite there. For as Benjamin reminds us, it is not so much the child *as a* child around whom these questions circle. It is the "image" of the child that provokes them: not the boy, but the carefully composed image of the boy in Quarmyne's series of photographs. Or, to rephrase things yet again, still slightly more precisely: it is the "figure" of the boy with which we are finally concerned; Quarmyne's stylized figuration of him as a kind of character; a type of character at once old and new; a historical and ultra-historical character; a representative character of the Anthropocene.

That is where my argument is headed.

But as the language of *figure* and *character* suggests, to fully grasp what is at stake in making it, we need to pause one last time before directly re-addressing the photographic image/character of Collins Kusietey on his Ghanaian shore to ask a similar set of questions of one or other of his literary doubles, one or other of those evidently novelistic "characters" who are his aesthetic cousins: his numerous and rapidly emerging literary kin scattered across the world-and-Eaarth-scene of the twenty-first-century Anthropocene novel. We will benefit from doing so not because the novel exceeds photography or the other arts in its significance to understanding the images of our time (or in anchoring the conviction, running throughout this text, that what we know of aesthetics, affects, and genres does indeed have a continued deep salience as we confront the crisis of climate change) but because in the practice and theory of the novel—particularly the historical novel, as that genre encounters the blended history and supra-history of the Anthropocene—we have access to a particularly significant grammar for comprehending and decoding these images, these image-characters, these representative "types" of our time.

With that in mind, let me pose the question again. Not yet of Collins Kusietey, but first, en route back to Quarmyne's image of him, of one of his most complex and representative fictional counterparts: Sonmi~451. What, now, can we know of her? Of the mixed epoch, heterogeneous circumstance, and the multilayered human and posthuman situation she sums up? Of the planetary conjuncture and order of time she represents (and reveals the image of Collins Kusietey also to characterize)?

Who is Sonmi~451? Why, within the context of this inquiry, should we concern ourselves with her? And what can we know of her?

To begin with, we know that she is a character in David Mitchell's novel *Cloud Atlas*—a "work," which, appearing at this point in my argument, carries the highly determined charge of "sublating" our mixed epoch within itself. If we are to care about Sonmi~451 (at least for the reasons I have outlined), it will be because in its characterization of her the novel proves itself capable of fulfilling that charge: not alone, not uniquely, but with sufficient constellating power to bring thought to arrest in the presence of one of those "images," which, for Benjamin, are the crystallizing zero-points of a historical "dialectics at a standstill." An image (of the epoch of the Anthropocene) such as this: "The ford cleared the urban canopy near Moon Tower, and I saw my first dawn over the Kangwon-Do Mountains. I cannot describe what I felt. The Immanent Chairman's one true sun, its molten lite, petro-clouds, His dome of sky."[46]

Why would this image throw thought into arrest? Why, arrested here, pondering the image of Sonmi~451's first dawn, might thought "explode a specific epoch out of the homogenous course of history"?

In part because of what the novel has already disclosed about Sonmi-451, the first-person "I" addressing us in this scene; in part because of what it will subsequently reveal about her.

That first dawn occurs approximately midway through the fifth of *Cloud Atlas*'s six narratives, which are sequentially embedded, matrushka-doll-style, inside one another, with the first and chronologically earliest narrative "beginning" and "ending" the novel; the next, and chronologically subsequent account, coming second and second-to-last; seriatim. The sixth narrative is thus the core of the novel. We know that Sonmi-451 is a fabricant, a genetically engineered "un-Souled" server designed for labor in a gleaming fast-food restaurant in mid-twenty-first-century Korea. She is not, thus, an individuated person with an individual name—or a name (Sonmi) and a surname (451)—but a model with a serial number, the 451st Sonmi "genomed" for a decade or more of servitude, alongside a set of "Yoonas" and "Ma-Leu-Das" and "Kyelims" in a fast-food camp and, thereafter, destined for the recycling of her genetic material. In Agamben's terms, Sonmi~451 is an emblem of pure "zoë," a model of "abandoned" life within a "Corpocratic" order, which her narrative ("An Orison of Sonmi~451") holds out as late, late globalization's new "nomos of the earth." Yet a curiosity of her physical being materially and obdurately complicates her exclusive assignation to this category of life.[47] She has a birthmark. And not just any birthmark: a birthmark in the shape of a comet that she shares with the protagonists of each of the four preceding narratives (set, in order, in 1850s Pacific, 1920s Europe, 1970s California, and 1980s England); a mark that, shared with these protagonists, sets her in a second order of kinship, not only with her utterly alike fellow Sonmi's in their mid-twenty-first-century present but with a set of ancestral relatives, cousins, and precursors through a biological (or biotic) line of descent processed through a mix of sexual and technological selections over a multi-hundred-year period of time.

Zoëlogically identical to all the other Sonmis of her time horizon, biologically linked to a vertical, multi-century line of bio-techno-evolution, Sonmi~451 nevertheless speaks in the first person, as an "I," as a conventional novelistic character, or, more precisely, as a conventional character in a realist novel. Which is to say that she speaks biographically, with all the assurances of liberal, individual personhood underpinning the modern, realist biographical form. If the oddness of that claim to the biographical by this character has not been apparent from the start of her account (formally structured as the transcript of an interview with a corporate "archivist"), then it is thrown into bright relief in the moment I have identified, the moment of Sonmi~451's "first dawn," the moment of enlightenment in which, emerging for the first time from the subterranean mall in which she has served the full course of her life, she sees what no other Sonmi has yet seen: a landscape, natural light, clouds, the sun. Recalling this dawn, unique to her, Sonmi~451 is in a significant way recalling her first fully biographical moment, her point of entry into biographical being, her initial passage into an individual life; a passage that will then encompass a Kantian labor of complete enlightenment, beginning with an intensive program of humanistic reading she plans to pursue at the university where she is taken to study (and be studied) and culminating in her complete rebellion against Corpocracy, her ensuing arrest, her prison interview with the archivist, and her execution.

Speaking to us about this first dawn, and so, in this sense, defiantly recollecting her birth as a biographical subject (rather than as a zoëlogical subject or a biological, or bioengineered, subject) from the aesthetic experience of this dawn, and, so, also, recalling her birth as a belated, humanistic subject— a subject of the conjoined enlightenment projects of aesthetic judgment and of freedom—Sonmi~451 thus maps quite an old story onto her passage into the future. She sets her account of a revolutionary movement into the possibility of emancipation within an entirely recognizable frame: the framework of freedom provided by the story of enlightenment progress; the story of educated cosmopolitan personhood; the story of normative citizenship, rights, and equal protection; the story of the self (of human life) at the heart of what Chakrabarty calls History 1 (and of what Lévi-Strauss highlighted as the core conception, and flaw, of Sartre's "historian's code" and its allied, materially humanist theory of freedom).

As *Cloud Atlas* reanimates that Enlightenment narrative of liberty, reviving the normative political imaginary of History 1, the striking image of the first dawn of Sonmi~451 springs its first arresting set of questions. What are we to make of this first-person, speaking "I"? Does she enter—in this way—the Enlightenment-humanist path of biographical life? Is she a recognizable "person" pivoting, in such an apparently conventional fashion, from subjugation, through the awakening experience of aesthetic astonishment, to the task of modern freedom?

Are we to celebrate her? Or should we mourn that she seems to deny that the truth of a person is multiple, that she seems to discard the ontological multiplicity of personhood, that she seems to choose the enlightened-biographical over the evolutionary-biological and the techno-zoëlogical orders of her being?

Or are those the wrong questions? Is the point, instead, not to assume that Sonmi~451 is choosing one or other of these multiple modes of being, but to look back at this image for the cues it provides to a nonexclusive form of relation between these plural temporal codes and orders of life?

If there is evidence in this image that Sonmi~451 does not simply revert to History 1 and the Enlightenment freedom of the autonomously, irreducibly, human self, but rather moves toward a conception of the person adequate to the situation of History 4°, it lies in the quality of the light pouring down on her in her quasi-Wordsworthian ("bliss was it in that dawn to be alive"), dawning moment. For while that light is quite clearly marked as sublime ("I cannot describe what I felt") and is, so, unmistakably linked to the sublime's distinctive history as an agent of Enlightenment humanization, neither it nor its sublimity, is figured in this image as "natural." Rather, the molten light from the Chairman's sun is filtered by petro-clouds. Looking up, to put it simply, Sonmi~451 does not see nature, but the collapse of human and natural history. Human history is projected into the heavens (a sky possessed and branded by a corporation) by clouds that are as humanly engineered as her own body. She sees parts-per-million particling the ether. She sees the world at 4°C. She sees the Anthropocene. Crucially, it is this sight of a human-refabricated nature, and not the flight from petro-light into pure organic light, that triggers for her an experience of the sublime. This new Anthropocene sublime, rather than being oriented to establishing a uniquely "human" capacity for transcendence in the mind's experience of nature (and so of affirming humanity's unique dignity apart from nature), is instead directed to marking a simultaneous capacity for immanence in the experience of the indistinction of the human and the natural, in the awareness of the blending (even the catastrophic blending) of the human and the nonhuman worlds. And it is this immanent rerouting of the transcendent that transforms the image of Sonmi~451 (at her first dawn, inheriting an Enlightenment project of freedom and then refracting it through the biographical, the biological, the zoëlogical, and the nomological orders and chronologies of her being) into an image capable of arresting thought and (in full Benjaminian form) of blasting our mixed Anthropocene epoch out of the homogeneous course of history.

And the theological? That final mode of the "extra-historical" nonhuman? It is not yet apparent, but it finally emerges, in the ensuing sixth (and innermost) narrative, set a century or more later in a post-apocalyptic world—an "ultra-historical" world, to borrow another of Agamben's terms—trapped in the long interregnum between the environmental ruin of the Corpocratic "final" stage of history and the establishment of a new post-ultimate reordering of the earth (or, as McKibben has it, the "Eaarth").[48] Within that world, saturated with extreme violence and a perpetual Hobbesian war of all against all, Sonmi~451 recurs as a figure of messianic possibility, not through any recollection of her revolt against her engineered "un-Souled" condition but as an object of sacred veneration. And that too gives thought pause.[49] For as Zachry, the protagonist of this narrative, recounts in oral form the adventures of his life and his encounters with a now-deified "Sonmi," *Cloud Atlas* seems to re-transcendentalize, or, more simply, to mystify her; to situate her within a decidedly non-immanent sphere of the "image."[50] The decisive moment comes at the end point of his account, as he remembers his own kayak-borne flight to freedom from a band of raiders who have killed his parents and siblings: "I watched clouds awobbly from the floor o' that kayak. Souls cross ages like clouds cross skies, an' tho' a cloud's shape nor hue nor size don't stay the same, it's still a cloud an' so is a soul. Who can say where the cloud's blowed from or who the soul'll be 'morrow? Only Sonmi the east an' the west an' the compass an' the atlas, yay, only the atlas o' clouds."[51]

One might think that Zachry's thought is indeed simply mystified, that as we know from the surrounding "Orison" narratives of the novel, there is nothing divine about Sonmi~451, that Zachry and the fellow members of his "Valleysmen" world who have come to regard her as a deity and have built a religion around her are merely engaged in a familiar anthropological category error: the sacralization of the secular. But the novel does not permit that option. For Zachry and his fellows, Sonmi is the source of auguries, cryptic but precise prophecies of what is about to unfold in their lives. And the difficulty is that all these auguries come true, one by one, exactly as prophesied. As that occurs, as Zachry's personal history is directly opened to the operation of prophetic miracle, the biographical scale of life that the earlier image of Sonmi~451's dawn revealed to be so intimately cross-layered with the biological, the nomological, the zoëlogical, and the geological is here repeatedly interrupted, patterned, modified by a theological force the novel resolves to treat as simultaneously inexplicable and real.

Zachry, in his turn, like Sonmi~451 looks up at the sky and sees not petro-clouds but some sacral-efficacious afterimage of the long-dead fabricant now appearing in the guise of "Sonmi the east an' the west an' the compass an' the atlas, yay . . . the atlas o' clouds." He sees, in fact, two things that correspond to the difference between what he sees and what becomes visible over his shoulder, so to speak. First, he sees an image of human life belonging not to a single but to multiple temporal orders: "Souls cross ages like clouds cross skies." As he sees this, the novel provides an answer of sorts to the conundrum of the repeating birthmark repeatedly borne by its characters across multiple narratives (and multiple narrative ages). That mark, we can now posit, allows all these characters to relate to one another as reiterations, or quasi-reincarnations, of one repeating soul, crossing from age to age. Or we can posit this as long as we are looking ahead in unfolding chronological sequence from one moment (one age, one narrative) to the next. Looking backward, however, from the end point of the final narrative, from the lag-end of Zachry's yarn, that relation reemerges as no longer prophetically reincarnational but retrospectively cumulative, as each life, each moment, each age gathers and contains within itself the ages and times and marks that have preceded it. Regarded from this perspective, the "soul" or spirit of any age (to speak in modified Hegelian terms) is not homogeneous and unitary, but multiple and layered. And that, of course, is one of the keys to Benjamin's revolutionary historical method, to his determination to resist the narrative of historical progress by altering perspective, by looking back over the shoulder of his "angel of history," at the shards of ruin and sedimented layers of history piling up at the angel's feet. Benjamin's messianic perspective (the messianic perspective of a historical materialism, even the messianic historical materialism of an extra-historical time) thereby reveals itself not as a perspective toward the prophesied future—toward the not-yet coming of Messiah—but toward the retroactive realization of a heretofore missed but still "revolutionary chance in the struggle for [a] . . . suppressed past": a past which is, in fact, neither "past" nor singular but multiply immanent in and marked on the present. This "civilizational" history is the species history of homo sapiens, the history of organic life on earth, the history of the universe and the blended orders of biographical, nomological, biological, zoölogical, and geological time.

This retrospectively apparent, cumulative nature of time is one of the things *Cloud Atlas* renders visible in Zachry's inspection of the age-crossing clouds "awobbly" above him. It also explains why, at this point of its furthest chronological progression, the novel then immediately pivots back in time, peering hinterward, one narrative strand after another, to hunt for clues of what has made Zachry's long Anthropocene ruinous present what it is, and to discover what (in the past ages drifting toward the time-accumulating present) might be reopened for revolutionary, emancipatory possibility. Yet it still does not touch what Zachry himself sees as he looks up into the sky, up into the clouds, and sees Sonmi there as "the east an' the west an' the compass an' the atlas"—and is miraculously rewarded for seeing her so (and believing in seeing her so). His vision here is the precise complement and obverse of what Sonmi~451 has seen in her earlier look skyward. It is not the collapse of the distinction between human and natural history in an apparition of petro-clouds, but the collapse of the distinction between human and sacred history in the apparition of Sonmi as atlas of those clouds. What Zachry sees (and what the novel, in respecting the efficacious power of Sonmi's auguries also agrees to see, if not to explain) is the blending, as Chakrabarty would say, of "the time of history" and "the times of gods."[52] Like his fellow Valleymen, he sees History 2, immanently full with the force, the plenitude, the possibility of a sacredly nonhuman order of actants, a distinctive, affective, vernacular way of being within the blasted totality of planetary catastrophe, a theological/cosmological alternative to a unitary Enlightenment passage into freedom within our looming moment of danger.

And as that recollection of *Cloud Atlas*'s fictional status highlights, it underscores that any method of thought adequate to encompassing this heterogeneous, epochally mixed moment must include in its appraisals of the collapsed distinction of human, natural, and sacred histories, and in its multiple encounters with the nonhuman, an encounter with the aesthetic, and an appraisal of the "actant" power of aesthetic history.

Sartre, as I have noted, prefaces *The Family Idiot* this way: "[It] is the sequel to *Search for a Method*. Its subject: what, at this point in time, can we know about a man? It seemed to me that this question could only be answered by studying a specific case. What do we know, for example, about Gustave Flaubert?" I have been echoing that query, and echoing the question behind it, asking what, at this point in time, in the epoch of climate change, we can know of the relationship between the domain of human action and the domain of historical necessity, the realms of making and of circumstance, of the actor and the "situation." As I have done so, I have both followed Sartre's lead and diverted from it, rereading Sartre's *Search for a Method* through the critique of his contemporary antagonist Claude Lévi-Strauss and a preceding and succeeding body of work (from Walter Benjamin to Dipesh Chakrabarty), calling to our attention the various and multiple ways in which the long-standing distinctions between human, natural, and sacred histories have fallen into collapse, and foregrounding the infra-historical and suprahistorical challenges to the "historian's code" Lévi-Strauss found underpinning Sartre's thought. In doing so, my concern has been, above all, to trace a new search for method emerging from a twentieth- through early twenty-first-century extra-historical dispositif (ranging from Benjamin's messianic historical materialism, to Lévi-Strauss's "savage" dialectic, to the neuroscientific and ecological turns of contemporary philosophical thought) and to identify the recent critical discourses on the Anthropocene (particularly Chakrabarty's contributions) as both belonging to that dispositif and offering to resolve its methodological search.

While I have thus based my discussion on an expanded variation of the question of freedom that Sartre inherits from Marx (the question of how we might find some method to make something of what our mixed, multi-scaled situation is making of us), I have also proceeded on the basis of a fundamental difference. Where Sartre's method rests content with the "universality" of "human history," I have been arguing that we must also address the infra- and supra-historical domains of planetary time. There is another difference as well: where Sartre ultimately brings his search for a method to bear, for test, on the life of a nineteenth-century novelist, I have turned for my own first test "situation" to a twenty-first-century novelistic character.

In closing this section I want to pause to ask what difference that aesthetic turn makes, what openings it affords, what it means to derive a method of accounting for the "situation" not from a real but from an imagined life, not from an "actual" person but from a "fictive" persona, not from someone like Gustave Flaubert but from someone like *Cloud Atlas*'s Sonmi~451.

How does Sonmi/Sonmi~451 differ from Flaubert? She is, to begin with, and to repeat the most obvious point, something Flaubert never entirely was: a character. It is, of course, one of the insights of Mitchell's novel that the past is always dually composed of an "actual" and a "virtual" past, the past "as it actually occurred" and the past "created from reworked memories, papers, hearsay, fiction—in short, belief—[which] grows ever truer" as the actual past "descends into obscurity." While we might say, in consequence, that to the "actual" past Flaubert, there are multiple circulating "virtual" Flauberts created from reworked memories, papers, hearsay, and fiction, and, so, multiple ways in which he too has become a "character" of our thought, Sonmi~451 has nevertheless always, only, been a character. Yet maybe this difference is not quite so obvious. For as we have seen, it is one of the conceits of the novel that Sonmi~451 also undergoes this transformation. In her passage from Sonmi~451 to Sonmi "the atlas of clouds," she, too, is not ontologically singular. To the later characters within the novel's extended landscape of time, she also shifts from what she "actually" was (a genetically engineered server) to become a character, a reworked fiction, an article of belief. And it is precisely in becoming such a character in Zachry's world that she becomes an actant within it (whether through the never-explained power of her auguries, or simply by becoming a guiding fiction, an atlas, for his actions regardless of the irresolvable question of those auguries' miraculous, history-interrupting force). In staging Sonmi~451/Sonmi so, in underlining her double life as a "person" and a "character" both for the readers of the novel and within the world of the novel, *Cloud Atlas* reminds us of something fairly old and fairly elementary, but worth recollecting nevertheless: the actual and the virtual are not strictly divisible, but entangled. Both "act" within the real, and just as "Sonmi" can function as a shaping fiction and a virtual sharer in the "actual" lives of her fictional acolytes, so too as a character (as "Sonmi" or "Sonmi~451") can she (or any other character) act in our "actual" space as readers of fiction.

Or, to put it in the terms of one of the critical discourses animating much recent theoretical engagement with the Anthropocene and much of my thinking here, the novel suggests that to the world of strange-strangers, nonhuman co-presences and co-actants in the extra-historical fields of human-history, we may need to add not just Chakrabarty's Gods, Tim Morton's hyper-objects, Jane Bennett's vibrant matter, the heat waves, Keeling curves, cyclones, melting ice, and carbon parts-per-million of climate science, but also "characters," "fictional lives," aesthetic phenomena; that if we are to unravel the ontological and temporal strands of our plural situation, we must begin with "the smallest possible unit of analysis," and that smallest unit, as Donna Haraway's *The Companion Species Manifesto* insists, is not the biographical unit but the "relation." To the relation of the biographical to the biological, the zoölogical, the geological, the cosmological, and the theological (all of which we have already seen Sonmi~451 trafficking), we thus also need to add our relation to these other companion-species of the human, these fictions, these "characters."[53]

But why would this matter? What difference might it make to add these aesthetic figures to a History 4°'s "parliament of things"?

I have two preliminary answers. One relates to the capacity of such characters to compose (or to allow us to see the composition of) a total world, and the other relates to their simultaneous capacity to decompose us and, in doing so, to recompose our conception of freedom. On the composing capacities of character, I have in mind, as might be predicted, characters of a certain sort, the protagonist characters of the historical novel as (or almost as) Georg Lukács understood them: the characters who serve as summarizing types of a particular moment or historical situation; characters who "embody the fundamental social and historical forces of their time," and are "located in a specific time and place and acting at crucial junctures of history . . . are who they are as a result of historical forces." They "represent the spirit of the age in their ability to inhabit the whole world in its totality, moving in and out of its separate spheres" (as Farhad B. Idris admirably summarizes).[54] Sonmi~451 is such a type, or very nearly so. For if these "historical forces" must now also be understood to include a full range of infra- and supra-historical forces (or as climate discourse has it, forcings), then she also represents the coming into existence of a new type of "type," a new form of the historical novel (perhaps the "extra-historical novel"), and, to borrow one more term from Lukács, a new "moving center" at the heart of this novelistic form.

As is well known of Lukács's view, the central charge of the realist tradition of the novel that incorporates (and in a fundamental sense begins with) the historical novels of Scott and Balzac is to map onto itself the "totality of objects" composing a historically particular social world, to demonstrate how "everything is linked up with everything else . . . [how] every phenomenon shows the polyphony of many components, the intertwinement of the individual and the social, of the physical and the psychical, of private interest and public affairs."[55] Rather than abandoning themselves, their readers or their characters to that polyphonic, intertwined, and potentially infinite "total" world, the novels of Scott, Balzac, and Tolstoy, Lukács insists, orient their maps of totality by discerning a social center of gravity around which all their linked objects and all their representative human types orbit, disposing and directing the movements of the "totality of objects." Lukács's term for this is "the moving center."[56] In depicting or grasping the nature of this moving center, the historical novel not only maps a totality, it reveals the often invisible force holding sovereign sway over that totality: the force by which it is unapparently but definitively animated. Where on Lukács's account, however, that moving center everywhere present within—and everywhere immanently sovereign over—the totality of objects remains resolutely within the realm of human history (as "capital" for Balzac, or "the peasant revolt which lasted from 1861–1905" for Tolstoy), in Mitchell's *Cloud Atlas* "this visible-invisible ever-present protagonist" emerges as another form of moving center, as another sort of visible-invisible forcing force exercising its sovereignty over a social totality: a forcing force crossing the border of human and nonhuman history, a forcing force expressing the planetary sovereignty of melting ice, temperatures Celsius, hydrologic cycles, "petro-clouds."[57]

And to render this forcing power, this moving center, this new global sovereign visible not only over the weather but over "the whole world in its totality," in all its "spheres," we need a new type, simultaneously expressive of Lukács's humanist Marxism and capable of surpassing it; a type of character capable of accumulating compound historical and "extra-historical" spheres, layers, scales, "ages," and forces; a type of character such as Sonmi~451 (and, as I will indicate, a type such as Nyani Quarmyne's Collins Kusietey); a character who in Chakrabarty's terms allows us

> to view the human simultaneously on contradictory registers: as a geophysical force and as a political agent, as a bearer of rights and as author of actions; subject to both the stochastic forces of nature (being itself one such force collectively) and open to the contingency of individual human experience; belonging at once to differently-scaled histories of the planet, of life and species, and of human societies.[58]

As Chakrabarty also makes evident in a later essay ("Postcolonial Studies and the Challenge of Climate Change," a follow-up piece to "The Climate of History: Four Theses"), the effect of being given this new "view" of the human, is not only (as it was for Lukács) revelatory of the totality of which the human situation is composed, it is also radically decomposing, radically decompositional of the image of the human and the portrait of history even a resiliently "anti-historicist" Marxist historical materialism had previously rendered apprehensible. And that is because, as he now indicates (in ways that "The Climate of History" less fully addressed), to gain such an image of the human is to gain an image of human life spanning three modes of existence: two of which possess long-standing provenances and relative ontological stability, and one that is utterly outside the bounds of previous forms of subjectivity.

Within the time of the Anthropocene, there is, first, the mode attending "the universalist-Enlightenment view of the human as potentially the same everywhere, the subject with the capacity to bear and exercise rights."[59] This is the universalizing mode of being postulated by History 1, the mode of being in which humans "are still concerned with justice even when they know that perfect justice is never to be had." It is the most recent iteration of a cosmopolitan-republican way of being for which the new arena of "climate justice" remains, as of old, fundamentally oriented toward the protection of the rights-bearing individual and securing the freedom of that individual from harm.[60] Then, Chakrabarty observes, there is a second mode of being attending "the postcolonial-postmodern view of the human as the same but endowed everywhere with what some scholars call 'anthropological difference'—differences of class, sexuality, gender, history, and so on."[61] This is the reappearing mode of History 2; the mode of humanity's vernacular differences; the mode in which (as "the uneven impacts of climate change" are "routed through all our 'anthropological differences'") justice finds itself oriented toward the preservation of collectively differentiated forms of life (particularly toward the protection of subaltern life-forms exposed, in globally asymmetrical patterns of vulnerability, to ecological devastation and geological change).[62]

But then Chakrabarty adds a third mode of humanity's being, a new mode into which we are now entering, a radically decomposing mode for which we have, he indicates, no model of "experience":

> [We] cannot ever experience ourselves as a geophysical force—though we now know that this is one of the modes of our collective existence. We cannot send somebody out to experience in an unmediated manner this "force" on our behalf (as distinct from experiencing the impact of it mediated by other direct experiences—of floods, storms, or earthquakes for example). This nonhuman, forcelike mode of ex-

istence of the human tells us that we are no longer simply a form of life that is endowed with a sense of ontology. Humans have a sense of ontic belonging. That is undeniable.... But in becoming a geophysical force on the planet, we have also developed a form of collective existence that has no ontological dimension. Our thinking about ourselves now stretches our capacity for interpretive understanding. We need nonontological ways of thinking the human.[63]

This is the distinctive mode of being within what I earlier called History 3; the mode in which, leaving behind the Enlightenment philosophical and political frameworks of citizenship, norms, and rights—and the subaltern frameworks of anthropological difference and "forms-of-life"—the question of the human (and of justice) passes into the realm of ontological dispute and renovation. It passes beyond ontological fixity, into what Chakrabarty calls the nonontological, bursting the relatively composed domains of citizenship and anthropological difference into a radically decomposed form of being. And here, the questions of justice and of freedom become much more vexed.

Climate scientists' history reminds us . . . that we now also have a mode of existence in which we—collectively and as a geophysical force and in ways we cannot experience ourselves—are "indifferent" or "neutral" (I do not mean these as mental or experienced states) to questions of intrahuman justice. We have run up against our own limits as it were. It is true that as beings for whom the question of Being is an eternal question, we will always be concerned about justice. But if we, collectively, have also become a geophysical force, then we also have a collective mode of existence that is justice-blind. Call that mode of being a "species" or something else, but it has no ontology, it is beyond biology, and it acts as a limit to what we also are in the ontological mode.[64]

At the conclusion of his second thesis in "The Climate of History," Chakrabarty points out that "the question of human freedom" lies open, exposed, and unresolved, "under the cloud of the Anthropocene."[65] In this later essay he repeats that point. Unable to experience ourselves in the new "mode" of radically decomposed, "nonontological" being we have taken on, we cannot conceive a new theory of justice or a new form of freedom adequate to that mode.

Two unanswered questions remain. Is it the case that we have no experience of such a decomposed "nonontological" way of being human? And if, alternately, we do—if we have some models for experiencing ourselves in this new mode of being—do such models point to a new way of conceiving justice (and a new practice of freedom) for such a nonontological order of life?

Before responding, let me offer one clarification. Throughout the argument I have been trying to develop, I have emphasized the need for uncovering a method of addressing not History 3 but what I have been calling History 4°. While History 4° encompasses History 3, it encompasses more than this, more than a nonontological way of being. In much the same fashion that for Chakrabarty, in his earlier work (most significantly in *Provincializing Europe*), an Enlightenment discourse of citizenship and of rights remained simultaneously "indispensable and inadequate" for a subaltern, postcolonial project of justice, so too, on my account, do anthropologically differentiated subaltern ways of being and the normative modes-of-being of cosmopolitan citizenship remain "inadequate and indispensable" to a full conception of human being in the mixed-epochal situation of this time.[66] Like any "situation," to put it another way, History 4° is not ontologically singular. It encompasses and blends multiple orders of time; multiple, allied, forms of being; multiple corresponding political imaginaries. It knots together History 1, History 2, and History 3. That is what I understand by History 4°. Chakrabarty comes very close to making the same point in "Postcolonial Studies and the Challenge of Climate Change," considering the "three images of the human." "These views of the human do not supersede one another," he writes. "One cannot put them along a continuum of progress. No one view is rendered invalid by the other." But then, importantly, he pivots away: "They are simply disjunctive. Any effort to contemplate the human today . . . on political and ethical registers encounters the necessity of thinking disjunctively about the human, through moves that in their simultaneity appear contradictory."[67] What I have wished to insist on (in a sense, by reading Chakrabarty against and through himself, by reading this most recent Chakrabarty through the earlier Chakrabarty of *Provincializing Europe*) is that just as History 1 and History 2 can be seen to be braided together in constantly inter-modifying ways, so too are the three human conditions of History 1, 2, and 3 knotted together.

The revised question, therefore, is not whether we have access to some form of "experiencing" ourselves (of thinking the human in the age of the

Anthropocene) solely as "nonontological," but whether we have some avenue for experiencing ourselves non-disjunctively across those plural forms-of-being collectively constituting the situation and the problem of being in our times.

The answer, quite simply, is yes, we do. We find it at Haraway's "smallest possible unit of analysis," at the level of the "relation"—in this case, at the level of a relation to a representational phenomenon, at the level of an aesthetic experience, at the level of a relation to a "character" such as Sonmi/Sonmi~451, a compound "character" compounding within herself (and within the circulating fictions of who she is) an aspiration for rights and protected citizenship, a capacity to constitute affectively rich and distinctive "forms of life," and the consequences of having made of ourselves a "geophysical" force of nature.[68] Experiencing our compound ways of being human in the aesthetic experience of a relation to such a character, we are also, then, given an image of a third form of justice and a third mode of freedom: a form of justice that does not begin and end with the urge to self-protection (individually or collectively) but braids into these an experience of being undone; an experience of being decomposed and recomposed through an entangling set of relationships to the biological, and the zoëlogical, and the geological, cosmological, and theological orders and times of planetary life; an experience that reorients the demands of justice from a unitary desire for immunity from these orders and forces; from a singular desire to be protected from them, to be free from them, toward a supplementary determination to refashion the biographical and nomological orders of our lives (our habits of dwelling, consuming, legislating) in relation to these other forces and forcings of planetary life—at this time and for times to come. From freedom "from" to freedom "toward"—that, as I understand it, is one new project for the future of our situation, one orientation for making something new of what we have been made, one new disposition toward the future.

But in order to grasp what this might look like—to determine how the challenge of freedom in the age of the Anthropocene (the challenge of a "freedom toward") does not entail leaving behind a discourse of rights and its negative liberties (its protections of vulnerable, disposable, abandoned life) but involves, instead, the work of braiding or blending together these seemingly contradictory freedoms, of imagining, and experiencing them together, across the historical and extra-historical scales and frameworks of our times—there is one more step that needs to be taken.

I have been arguing that the "situation" of the Anthropocene comprises an internally heterogeneous "totality." Much of my effort thus far has been to more clearly delineate the various elements of that totality by identifying the diverse temporal scales I understand to order the epochal time of the Anthropocene (a biographical scale, a nomological scale, a biological scale, a zoëlogical scale, a geological scale, a cosmological/theological scale). The problem now is to grasp the form of the relation of these scales to one another. If the Anthropocene, to put it another way, is to be grasped as a "totality" (as encompassing the "total" future history of the planet) but needs at the same time to be understood as internally plural (as having different conditions of visibility, differential distributions of vulnerability, differing claims on our understanding of justice when we approach it through the perspective of an individual human life, the political orders of metropole and postcolony, humanity in its "species" being, humanity as one species of life among others, etc.), then how can we understand the relation of those different phenomenologies, precarities, and political exigencies to one another? What form does their relation take? How can we grasp them within the structure of the "both/and"—particularly as they come to bear on the question of freedom?

By long-delayed way of answer, let me now return to my starting point, to that image of Collins Kusietey, standing in the "remains of a house destroyed by the encroaching sea in Totope, Ghana."[69] What, now, with the method of understanding I have articulated, can we say or know of him? What, in Benjamin's sense, does this image—this representation of the "dialectics" of forces and forcings at a "standstill"—reveal? To what orders of time, being, and politics does this image belong? What can it teach us of the task of freedom in the age of the Anthropocene?

3. The View from the Shore

What, now, can we know of Nyani Quarmyne's image of Collins Kusietey? Of the mixed epoch, heterogeneous circumstance, and multilayered human and post-human situation that image sums up? Of the planetary conjuncture, order of time, and orientation toward freedom it represents?

First, there is the name, the date, the age, and the place—all carefully noted in the caption: Collins Kusietey; March 7, 2010; seven years old; Totope, Ghana. Precise, identifiable, situating.

At what order or scale of meaning do these signs operate? In Lévi-Strauss's terms, the answer is quite evident. Taken at this place, on this day, Quarmyne's exquisite, aching image circulates at the "bottom" scale of Anthropocene time, at the level of the "biographical and anecdotal." If the genre of the Anthropocene were the genre of the diary, this image, read at this level of intelligibility, would constitute one of its daily entries. That would not lessen the affective and political powers it gathers within itself.

What powers? Those too are evident. First there is the power of sentimental identification on which many humanitarian projects have so long depended, a power of sympathy which Quarmyne's elaborate composition seems designed both to invoke and to query as the gilt-edged window-framing of the child tropes the propensity of the sympathetic view to bracket and contain the capacities of the lives (recurrently, the black lives) sympathy frames. Alongside that doubly summoned *and* abjured power, the image claims another, the power of sympathy's supposed opposite but secret double: the Enlightenment power of "the rights of man," rights held by this boy and threatened not only by the enduring history and legacy of the Atlantic slave trade visible in the cinder-block poverty of a village stretched along a parcel of Gold Coast shore scant miles from the lingering remains of a slave-trade "factory," but also, now, by a mounting rights-indifferent sea, a sea unaware and intolerant in its creeping rise of any claim this child may make to the right of health, or the right of shelter, as the future history of the Atlantic Ocean merges with the unfinished history of the Atlantic trade to leave Collins Kusietey trapped at the exact dialectical meeting point of forces and forcings I have been tracing.

Whether we elect to view him primarily as a subject of sympathy or as a rights-bearing subject as he confronts these twinned foes of his future (and his freedom to shape that future), the key point remains the same. Regarded from this perspective, from this first scale of analysis, this image of a single, named, and exorbitantly vulnerable citizen of the Anthropocene, captured on one dateable moment, at one identifiable place, emblematizes the enduring power of the biographical to frame and invite us into life and politics in the epoch of the Anthropocene very much as if we were still residents of Chakrabarty's progressive History 1.

To gesture, however briefly, to those histories of sentiment and rights is also, however, to begin to recognize the image's place within a second of Lévi-Strauss's scales of time: the scale I have been calling the nomological; the scale one "level" up from the individual; the scale, as Carl Schmitt describes, of socially organized inscriptions and orderings of the planet; the scale of time playing out not over days or weeks or years but over a bundle of decades, or a century, or more; the scale of what we are used to calling "periods."[1] To what nomological period of time does this image belong? Not one but many. I have already mentioned an Enlightenment and a Romantic-Sentimental nomos of the earth (or what, in the first volume of this series, I called the world-ordering views of "cosmopolitan interestedness" and "liberal cosmopolitanism" so central to the constitution of the modern).[2] Allied with and against these, wrapped in and around and oblique to them, are a series of others: to name but two (themselves deeply intertwined), the nomos of a Black Atlantic modernity, and the allied but more generalized nomos of the global postcolony.

This is not the place to review in any thoroughgoing way the overriding Black Atlantic and more broadly postcolonial orders of the modern world. That work has been comprehensively undertaken and continues to be pursued by generations of scholars on whose research and insight I have attempted to build. It *is*, however, the place to ask how our understanding of the postcolony and the Black Atlantic looks from this perspective, how they are cast into relief from this coast, from this angle of inspection, with the waters rising to the horizon line, pushing their waves, and surf, and bullions of beach sand through crumbling walls and doors in what only looks like an act of natural destruction but is, in fact, a historically natured (or alternatured) process of violence.³ To do so—and so to begin to grasp the work this image performs in rendering visible a dually human and alternatural "period" of history, an anthropogenically hybrid postcolonial and Black Atlantic chrono-nomos of the Eaarth—we need to return a final time to the material science of the Anthropocene.

To each of the Representative Concentration Pathways (RCPs) it outlines for the planet's passage into our four alternating climate futures, the IPCC's Fifth Assessment Report assigns a likely range of increase in global mean sea level rise (GMSL) by the year 2100 (relative to sea level averages between 1986 and 2005). For RCP 2.6 that increase in global mean sea level will be between 0.26 and 0.54 meters, for RCP 4.5 it will be between 0.32 and 0.62 meters, for RCP 6.0 it will be between 0.33 and 0.62 meters, while for RCP 8.5 it will be between 0.45 and 0.81 meters above current levels. As the report further notes, these are the more cautious of two sets of estimates, reached by using one modeling technique (the process model) rather than the other contending approach, known as semi-empirical modeling (which takes into account both observed data and records from prior geological periods such as the Pliocene, when CO_2 levels were similar to those we are approaching, and sea levels were significantly higher than at present). Under the semi-empirical approach, each of the RCPs would generate increases in ocean level "that are about twice as large as [those predicted by] the process based model" with the relatively moderate RCP 4.5, as one example, producing increases in sea level of up to 1.2 meters (greater, that is, than estimated for RCP 8.5 in the more conservative scenario).⁴ As the report further clarifies, regardless of which of those two approaches is correct, the full impact of climate change on sea levels can only be seen by looking beyond the year 2100 time-horizon and by turning from global to regional projections.

When the time frame of analysis is extended beyond the current century, "it is virtually certain" that under every pathway into the future, "global mean sea level rise will continue" to climb beyond the ranges of the 2100 estimates.⁵ By 2300, unless the concentration of carbon is held below 700 parts per million, the increase would not be between 26 and 81 centimeters (in the best-case scenarios for the twenty-first century) but between 1 and 3 meters. "If warming is sustained for several millennia," those amounts would not be 1 to 3 meters in total "but *1–3 meters per* [each] *degree of warm-*

ing."⁶ Even if the worst temperature increases are avoided, even if the planet somehow avoids its rendezvous with a 4°C world, the report ominously concludes, "the available evidence indicates that *global warming greater than a certain threshold* would lead to the near-complete loss of the Greenland Ice Sheet over a millennium or more, causing a global mean sea level rise of about 7 m[eters]."⁷ What is the threshold for that twenty-three-foot increase in global ocean levels? "Greater than 2°C," the report quietly notes, "but less than 4°C of global mean surface temperature."⁸ Unless that is, the estimates take "into account the increased vulnerability of the ice sheet as the surface elevation decreases due to the loss of ice . . . [in which case] the threshold could be as low as 1°C."⁹

On a moment's reflection, this is an utterly daunting finding, more so, perhaps, than any other of the report's individual conclusions. For if it is the case that with as little as a 1°C change in temperature, a millennium from now thermal expansion, glacial melt, and the collapse of the Greenland Ice Sheet will have lifted the globe's waters seven meters higher than they are today, then not only does the scale of damage attending the long coming of climate change leap exponentially higher than anything I have yet noted, but so too does the task of thought as it confronts and seeks to address or even counter, through its own "force," the advent of the Anthropocene.

This does not mean, however, that thought has time to wait. For well before the coming of this thirtieth century, well before the culmination of this massively extended process of "slow violence," the intense and life-devouring manifestations of this planetary future have already begun to arrive. If not yet persistently, then episodically. If not yet evenly across the full surface of the globe, then regionally—and with a distinctive quality to that regional unevenness. As a World Bank report following up on the initial *4°: Turn Down the Heat* document notes, it is not just that climate change advances both inexorably and asymmetrically, with significant regional variations in the expression of its consequences, but that in its uneven advance, the regional shadings of the warming planet map with an uncanny precision onto our older maps of metropole and colony, of "center" and "periphery," of the hegemonic and the subaltern. If that is true of virtually all the Anthropocene's catastrophic effects, it is particularly true with regard to sea level rise as the advance event-horizon of climate change—as is quite graphically revealed in one of the maps accompanying the AR5 working group report highlighting the regional variation in sea level rise over the immediate decadal and centennial scale of Anthropocene time and revealing where change will

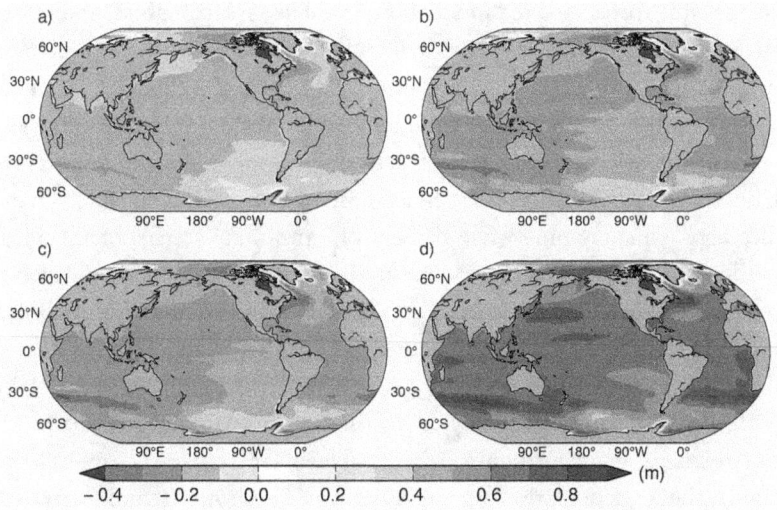

Figure 3.1. "2013: Sea Level Change." From *Climate Change 2013: The Physical Science Basis. Working Group I Contribution to the Fifth Assessment Report of the Intergovernmental Panel on Climate Change*, Cambridge University Press, 2013.

come most swiftly and most intensely within the next half century: along the Atlantic and Indian coasts of sub-Saharan Africa, portions of the North American mid-Atlantic, and in Asia and South Asia.

What will the advance of the rising seas mean in these places, and what will it betide across this new map of the Anthropocene postcolony? Among many others, it will mean that from 2005 to 2070 in five Southeast Asian cities alone (Jakarta, Yangon, Manila, Bangkok, and Ho Chi Minh City), the number of individuals extremely vulnerable to coastal flooding will leap from 3.9 million people to 28 million people. It will mean that in Bangkok, "without adaptation," by that same time almost 70 percent of "the area [of the city] is projected to be inundated due to flooding linked to extreme rainfall events and sea level rise"; that the Mekong Delta, which "produces around 50 percent of Vietnam's total agricultural production," could lose up to 12 percent of its crop production by 2040 "due to inundation and salinity intrusion."[10] It will mean that across Southeast Asian coastlines the "intensity and maximum wind speed of tropical cyclones making landfall are expected to increase significantly" before the century ends, with substantial attendant increases in loss of lives, shelter, and sanitation.[11] It will mean that Bangladesh, in particular, will emerge "as an impact hotspot with sea level

The View from the Shore · 77

rise causing threats to food production, livelihoods, urban areas, and infrastructure."[12] In sub-Saharan Africa, it will mean that by 2100 the number of people subject to annual flooding will double, from 9 million to 18 million (with 3 million Nigerians and 5 million Mozambiquans facing flooding every year). It will mean that infrastructure in coastal zones will be particularly "vulnerable," with ports such as those in Dar es Salaam, which "handles approximately ninety-five percent of Tanzania's international trade," perilously at risk.[13] It will mean that in the poorest continent on the globe, "economic damage induced by coastal flooding, forced migration, salinity intrusion, and loss of dry land" will reach $3.3 billion annually by 2100 and that, as the economic losses mount, access to safe drinking water will fall, with "threats . . . especially high along the Guinea coast and East Africa."[14]

And these, again, are only the predictable linear consequence of sea level rise in these particularly vulnerable African, South Asian, and Southeast Asian regions of the globe over the coming decades. The "cascading impacts" of intersecting changes, as the report summarizes, will mean that across these areas of the world, "deltaic regions and coastal cities" will be "particularly exposed to compounding climate risks resulting from the interacting effects of increased temperature, growing risks of river flooding, rising sea level and increasingly intense tropical cyclones, posing a high risk to areas with the largest shares of poor populations."[15]

As that final sentence indicates, the uneven global distribution of human vulnerability to the highest risks and worst consequences of sea level rise result from the interplay of two intersecting factors (what I have been calling the dialectic of forcings and of forces). First, the oceans rise unevenly because the radiative forcing of CO_2 emissions plays off a set of uneven conditions fundamental to the geology of the planet. "Because of the gradual melting process" attending temperature rise, the "gravitational pull" of polar "ice masses" is decreasing, thus accentuating "[sea level] rise in the tropics, far away from the ice sheets . . . [while] close to the main ice-melt sources (Greenland, Arctic Canada, Alaska, Patagonia, and Antarctica) crustal uplift and reduced attraction cause a below-average rise, and even a sea level fall in the very near-field of a mass source."[16] In combination with "ocean dynamics, such as ocean currents and wind patterns," that variation in gravitational pull is now driving the regionally uneven "pattern of projected sea level rise" traced by the report.[17] But neither this regional asymmetry in the consequences of heightened radiative forcing over the entire planet nor base topographic factors such as the low-lying "archipelagic landscape" of much of Southeast Asia alone can account for the uneven human vulnerability to the uneven physical distribution of ocean rise.[18] That vulnerability, instead, results additionally, and dialectically, from long histories of colonial underdevelopment and enduring histories of poverty in the world's postcolonies. As the report concludes:

> The poorest urban dwellers tend to be located in the most vulnerable areas, further placing them at risk of extreme weather events. Impacts occurring even far removed from urban areas can be felt in these communities. Food price increases following production shocks have the most deleterious repercussions within cities. The high exposure of poor people to the adverse effects of climate change implies the potential for increased inequalities within and across societies. It is as yet unclear how such an effect could be amplified at higher levels of warming [or at the higher and higher levels of rise that are due to arrive beyond the year 2100] and what this would mean for social stability.[19]

Faced with this amplifying and uneven vulnerability of human life before the present and future coming of the Anthropocene, this uncertain but catastrophic interplay of forces and of forcings, the view on the postcolony from the village of Totope turns magnifyingly bleak. Is humanly produced climate change solely responsible for Totope's vanishing shore? No, no more than climate change is the sole factor accounting for any of the storms, floods, and erosions of our times. But as its base topological conditions and its history of humanly produced vulnerability enter into exchange with all the forcing-forces of the Anthropocene, this single outpost reveals itself to be not alone but part of a global series, a series vaster and more sprawling even than the series constituting the slave-coast factory system, a series of coastal small towns and coastal megacities, a series of a multimillion-numbered, planetary, postcolonial citizenry collectively disclosed as inhabitants of the capitals of the Anthropocene world. As the planetary clarity and range of that view magnify, so too does the temporal horizon of the nomological scale of planetary time illumined by Quarmyne's art expand: stretching back through four centuries of Black Atlantic history and another millennium into the future.

Time does not pass, it accumulates. That was the central proposition of the theory of history I identified in writing the first volume of *Specters of the At-*

lantic, an account of the slave trading voyage of the *Zong* after Captain Luke Collingwood loaded his human "cargo" at Fort William in Anomabo, on the coastline with Totope, prior to setting sail for the Americas and determining to drown 132 of his captives in exchange for their insured, speculative value. There are other, stronger ways of saying this. "All of it is now, it is always now . . . even for you who never was there." This line from Toni Morrison's *Beloved* is perhaps the most powerful and the most definitional articulation of an African American and Black Atlantic theory, practice, experience, and art of damaged, traumatic history into which Quarmyne's work has now entered conversation.[20] Sharpe's *In the Wake*, with its striking commitment to navigating a "past that is not past," with its commitment to fashioning a "way-making tool" through the wake of the held lives, imprisoned lives, drowned lives, "afterlives," survival lives, need-to-survive lives of black lives, is one of the more achingly powerfully recent additions to that tradition.[21]

But as he adds to that tradition, his work also reveals a new dimension of what it means to regard time so. For as we cannot now *not* see, the vastly distributed Anthropocene postcolony metonymically figured by Totope and the other settlements pictured in his series accumulates more than the ruinous un-passed time of human history alone. It contains more than the time and the damage of the centuries-long slave trade and the overlapping/successive experience of colonial rule; more than the revolutionary accomplishments, post-independence frustrations, millennial aspirations, and aesthetic innovations of the era of the postcolonial; more than the legacies of decades of "underdevelopment" shaping post-independence Ghana's fragile systems of health care, education, housing, and urban infrastructure; more than the accumulated and flashingly brilliant arts of survival and creation of which "We Were Once Three Miles from the Sea" is itself such a powerful example.

As those images and the reports of the IPCC and World Bank reveal, the coastal postcolony also contains the accumulated consequences of carbon accumulating in the atmosphere over the past two centuries, warming the planet, raising the oceans both globally and unevenly. And as these works of aesthetic creation and empirical accounting reciprocally further reveal, those two cycles of accumulation have now entered into a devastating feedback loop with one another. Accumulated centuries of designed disposability and engineered precariousness have geomorphed with the accumulated (globally common *and* globally uneven) effects of climate change to render the postcolony singularly vulnerable to and singularly expressive of a now doubly *geo*political condition 400 years old and extending a millennium into the future: a condition merging the long-visible global politics of centers and peripheries with a newly visible geological exacerbation and intensification of that still unpassed and human-organized distribution of harms, vulnerabilities, and unfreedoms so fundamentally constitutive to the making of the modern world (and the modern's pre-organization of the 4°C world to come).

This is what Quarmyne's series reveals as the nomological code of the postcolony in the epoch of the Anthropocene. This is the second doubly accumulating, massively extended order of time that encompasses Collins Kusietey as he stands on the shore gazing at us.

Or, to add a proposition to my earlier contention: *time does not pass, it accumulates; accumulation does not end, it doubles back, and piles on again.*

But matters do not end there. For the more we abide with this image and the series of photographs of which it is a part, the more apprehensible is the multilayered complexity of the figure (or "character") at its center. The child, Collins Kusietey, like the matrushka doll image of Sonmi~451, increasingly discloses himself to represent a new type of "type": a type of Anthropocene person, a type of Anthropocene character indexed not only to a biographical and a nomological code (the classical materialist codes of the "actor" and the "situation," of a *particular life* lived in a *particular social and political world*, even as—in a departure from or innovation on classic materialist epistemologies—we have come to see that world as both historically and alternaturally co-constituted) but to a broader set of codes linking this child's life not only with the life-worlds and life-forms of the postcolony but with bios and zoë and geos—with the fate of the human as a species, with humanity's identity as one species among others in the realm of life itself, with the entire coming geological future of the planet.

How so? To detect this we must return to the image and series of photographs to which it belongs to note something markedly evident on which I have not yet remarked.

Viewed in full array alongside one another, the ten images of the series possess a singular power and force, reverberating back and forth from one scene to another: from the solitary pose of Collins Kusietey; to the worn visages of an elderly couple gazing out at us from the window of their shattered home; to the beach sand piled waist-high to a man virtually imprisoned in his crumbling house; to the haunting exposure of a mother and her children, effluent-thick storm waters pooling their feet as they stand bearing radiant

Figure 3.2. Collins Kusietey. From "We Were Once Three Miles from the Sea." © Nyani Quarmyne/Panos Pictures.

Figure 3.3. Anikor Adjawutor and Miyorhokpor Anikor. From "We Were Once Three Miles from the Sea." © Nyani Quarmyne/Panos Pictures.

cans of paraffin light like some desperate spark snatched against the coming of the Anthropocene night.[22]

The sequence possesses a singular force—one virtually definitional of Benjamin's epoch-defining images seized "at a moment of danger." But it possesses more than that power alone. That force is multiplied because, however singular its scenes of devastation, Quarmyne's series belongs to a broader, amplifying set of images—indeed to two broader sets, which it brings together. One of those is highly contemporary; the other dates back further in time.[23] But they have something in common. They are both series of artworks (across media: film, fiction, poetry, photography, eco-acoustics, installation) animated as arts of submergence, of drowning, of the underwater. The more contemporary of these, a set of works readily identifiable as art produced in the moment of and in response to climate change, include Paolo Bacigalupi's novel *The Windup Girl*, with its depictions of an ocean-threatened, drowning Bangkok; the flooding Sundarbans of Amitav Ghosh's *The Hungry Tide*; the dystopic coast-devoured South of Omar El Akkad's *American War*; the submarine Manhattan of Kim Stanley Robinson's

Figure 3.4. Numour Puplampo. From "We Were Once Three Miles from the Sea." © Nyani Quarmyne/Panos Pictures.

New York 2140; the submerged, coral-becoming underwater sculptures of Jason deCaires Taylor; and the ice-sheet-calving, levee-bursting floodscapes of *Beasts of the Southern Wild*.

Taylor's sculptures and Benh Zeitlin's film *Beasts of the Southern Wild* belong, simultaneously, to the second series of works I have in mind, for which the image of the ocean-abandoned black body is emblematic. The racial coding of the bodies of the water-threatened and the drowned are self-evident in *Beasts of the Southern Wild* but more nuanced in Taylor's underwater sculptures, though the iconography of the drowned slave crucial to so many of these pieces is difficult to miss. Another work that sits comfortably in both series is the Turner Prize–winning Otolith Group's *Hydra Decapita*, a series of video-installation pieces animated as a meditation on "the subaquatic descendants of Africans drowned by slavers during the middle passage"; and of course, Quarmyne's own "We Were Once Three Miles from the Sea," features both themes.

Other works that belong solely to this second set include the submarine cemetery scenes of Derek Walcott's *Omeros*; the underwater abyss that opens

Figure 3.5. A Family in Azizanya. From "We Were Once Three Miles from the Sea." © Nyani Quarmyne/Panos Pictures.

Edouard Glissant's *Poetics of Relation*; David Dabydeen's *Turner*; M. Nourbese Philip's *Zong!*; Fred D'Aguiar's *Feeding the Ghosts*; Morrison's *Beloved*, and countless others.[24]

In *Specters of the Atlantic: Volume I*, I argued that this image of the black body disposed to the waves has become an—perhaps *the*—enduring, repeating, time-accumulating image of the Black Atlantic. Now, at the writing of this volume, it is clear that that same image is also flickering into the sight-horizons of the world scene of Anthropocene art; defining the looming planetary condition of all human life abandoned to the power of the rising waves; illuminating the planetary condition of the entire human species in the epoch of the Anthropocene. In such works—certainly in "We Were Once Three Miles from the Sea"—we see those two imaginaries (the imaginary of the Black Atlantic and the imaginary of the Anthropocene) coming together.

There are multiple implications of this union, three of which I wish to stress. The first is that here, once again, we see that movement I earlier commented on—the movement from the constrained particularities of the

Figure 3.6. 2007 Vicissitudes, ID 3960. © Jason Decaires Taylor. All Rights Reserved, DACS/ARS 2019.

located subaltern (the residents of Totope on the coast of Ghana; the black poor of New Orleans under assault by Katrina; the black poor of Houston, a decade later, when Hurricane Harvey hit) to the generalized unfreedom of the species; as we also see the movement from the juridical and institutional domain of the citizen (as a subject of law, rights, and socially organized forms of life) to the ontological domain of the human; and the movement from a discourse on justice and freedom to a discourse on extinction—patterning the scale-jumping, scale-blending heterogeny of an Anthropocene totality.[25]

The second is that we now certainly have strong warrant to return again to a text such as Jean and John Comaroff's *Theory from the South: How Euro-America Is Evolving toward Africa*, both to revise and scale up its core argumentative claim (now better understood in terms of recognizing our moment as one in which the planet is evolving toward the Black Atlantic) and

Figure 3.7. 2007 Vicissitudes, ID 5075. © Jason Decaires Taylor. All Rights Reserved, DACS/ARS 2019.

to join the Comaroffs, together with scholars like Achille Mbembe, Sarah Nuttall, and Ato Quayson, in investigating how the practices of flourishing and survival nourished in the postcolony may increasingly, of necessity, come to inform the planetary practices of the species.[26]

The third implication is this: if it is the case that in the age of the Anthropocene the entire planet is in process of evolving toward the experience and the worldview of the Black Atlantic, that the future of the species is tending toward an encounter with the sovereignty of the rising oceans, the terror of the submarine, and a now fully global experience of Glissant's transversal abyss, then in recognizing this possibility—latent and manifest in Quarmyne's images, quantified and warned against in the IPCC and World Bank reports, devastatingly coming ashore, year after year, from Haiyan to Harvey to Irma to storms not yet named but inevitable—we see more than the third scale of time this black Anthropocene art shimmers into light,

The View from the Shore · 87

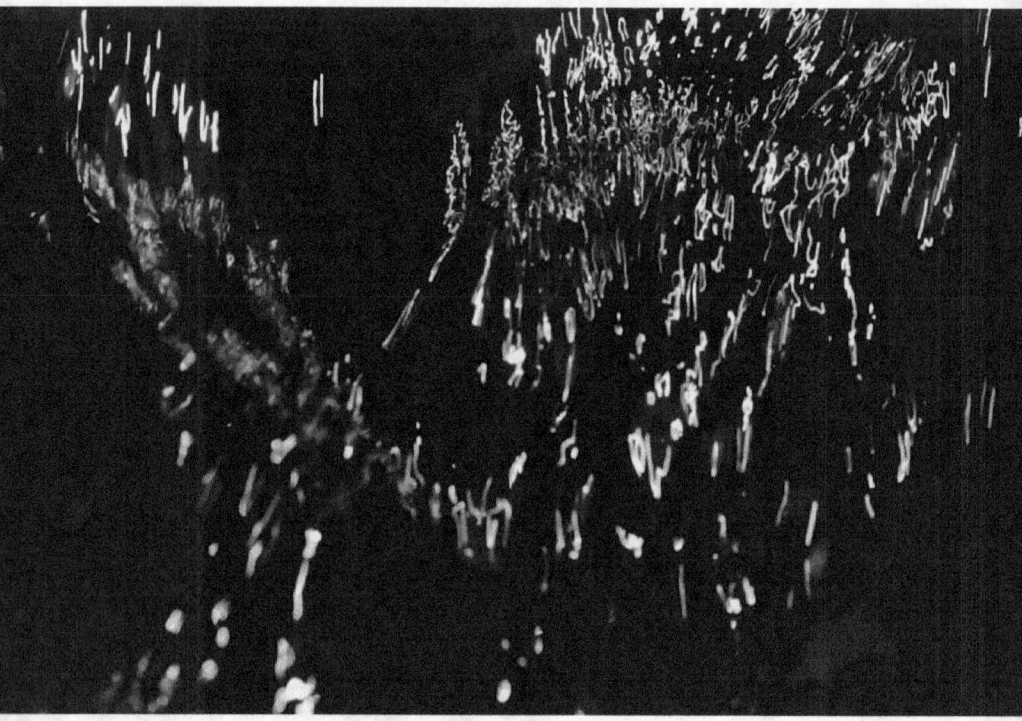

Figure 3.8. Still from Otolith Group. From *Hydra Decapita*. Courtesy of the Otolith Group and Lux, London.

more than the multi-millennial bio/geological scale of humanity's arriving planetary species life. We also come to see the little boy at the heart of Quarmyne's series as more than a single boy. Instead—as the title of the series suggests—we see him as a child who contains a planetary "we." And as we see this, we see one more thing, something crucial to the argument I have been advancing, something central to my claim that in the time of climate change, what we have known of the colonial and the postcolonial, of subalternality and hegemony, of contact zones and spaces-of-flow, do not sit outside the challenge of coming to terms with our new "planetary conjuncture" but *define* it.

We see that *we* come to the planetary *through* the subaltern; that we come to the "we" of the Anthropocene from the postcolony's shore.

With that claim, we return directly to an encounter with one of Chakrabarty's central arguments regarding the ontological disruption and becoming-

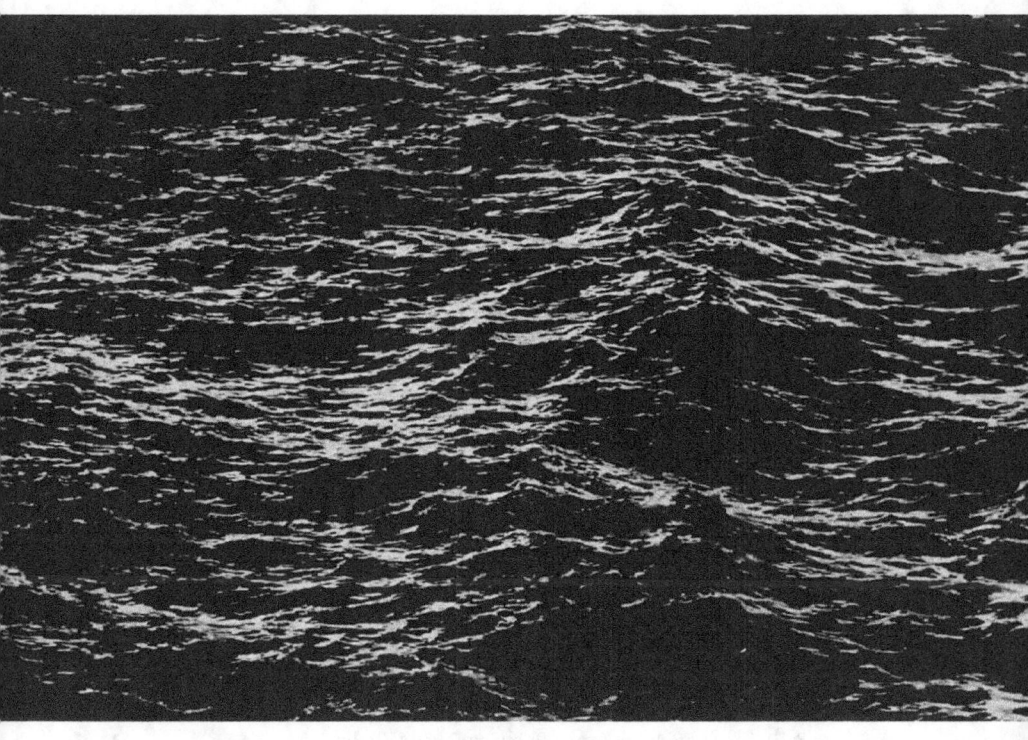

Figure 3.9. Still from Otolith Group. From *Hydra Decapita*. Courtesy of the Otolith Group and Lux, London.

species of the human in the time of climate change, the argument that there is, in fact, no access to such a "we." To be sure, as I have already noted, Chakrabarty qualifies that claim in ways that might seem to render the argument I am making esoteric to the point he has advanced. For when he insists that "[we] cannot ever experience ourselves as a geophysical force—though we now *know* that this is one of the modes of our collective existence," he goes on to clarify that what he thereby intends is that "[w]e cannot send somebody out to experience in an unmediated manner this 'force' on our behalf (as distinct from experiencing the impact of it mediated by other direct experiences—of floods, storms, or earthquakes for example)."[27] If the "we" of Quarmyne's title was a "we" (even across the range of the "we" of the biographical, the "we" of the sociopolitical, and the "we" of the species) that only experienced the "*impact*" (through floods, storms, and ocean rise) of the "geophysical force" of climate change, rather than a "we" revealed as

Figure 3.10. Houston, August 2017. Photo by David J. Phillip/AP.

co-being that force, then the "we" of "We Were Once Three Miles from the Sea" would have nothing to say to Chakrabarty's claim.

It is precisely for that reason that that claim must be taken so seriously.

For the claim demands to know what we do now see when we look at these images of Collins Kusietey and his fellow citizens of Totope. It demands to know whether we see them only being acted *on* by that menacing sea, or whether, somehow and simultaneously, we also see them (and thereby ourselves, whoever and wherever we are) acting *with*, *in concert* with, and *as* those waters; whether in seeing these images we can see, apprehend, and—through that smallest unit of aesthetic relation I sought earlier to tease out—"experience" a "we" not only victim to this geophysical force but party to it, part with it, particled in and among and *as* that force.

At the register that is core to Chakrabarty's project, seeing so means committing ourselves to a process of Anthropocene seeing—and reading, and understanding—after innocence. It means committing to an epistemology after and beyond an epistemology predicated on distributing differential responsibility for the damage of our times; committing to a reading of the human species as no longer meaningfully divisible between those who are and are not culpable *for* the force we have become but, instead, as humanly unified *by* the ontological (and ontology-shattering) shock of collectively *having become* the force we have become. At another, more immediate level, it means re-grasping the import of Quarmyne's title, understanding that the vanished separation it encodes, with its "we" and its "once" and its "miles" and its "sea," registers not merely an annulled separation of physical distance—"*we were once* three *miles*"—but also an obliterated ontological distinction between the "we" and the "sea." If it is the founding predicate of critical Anthropocene thought that the distinction between human and natural history has collapsed, then the title of Quarmyne's series requires more than a literal reading of the loss of yardstick distance between human settlements and the rising sea. It demands, instead, a reading of a human "we" who/which might have once understood its/our ontological terrain to be miles distant from the fabric of the ocean—radically disjunct from the fabric of ocean matter—but can no longer do so.

Or to put it one further way still, reading Quarmyne's "We Were Once Three Miles from the Sea" through Chakrabarty's postulate challenges this:

- Do these images disclose two ontologically discrete forms of matter inimically lined up against one another?
- Human matter against ocean matter; matters of history against matters of nature?
- Or do they reveal a world now composed of one blended, posthuman, material form?

As those questions ask us to answer what matter of matter we are now dealing with, they also ask that we return to the long-deferred investigation I have promised of the distinction, and the relation, between Materialism I and Materialism II.

I have argued that the young boy, Collins Kusietey, the central figure (or "character") of Nyani Quarmyne's "We Were Once Three Miles from the Sea," like David Mitchell's Sonmi~451, emblematizes the new "type" of the *type* of the Anthropocene world. He typifies and reveals an order of personhood bringing together (and simultaneously embedded in) multiple scales of Anthropocene time: scales of the biographical, the nomological, the biological, the zoëlogical, the geological, and the cosmo/theological. I have further suggested that in doing so, in standing at the overlapping fringe of all these orders of time, he—like Sonmi~451—renders visible a "we": a collective human species itself composed of all these modes of time and being coming into visibility not as an abstraction but through the shattering spotlight shone on his/this life lived on the Black Atlantic shoreline of the Anthropocene postcolony.

Here I echo and extend one of Gilroy's most celebrated arguments: if the women, men, and children of the Black Atlantic were the first truly "modern" people, so too, now, are the most precarious citizens of the postcolonial Black Atlantic (and the broader coastal and deltaic planetary postcolony they figure) the first truly "Anthropocene" people—the avatars of a fully globe-encompassing climate-changed world to come. But there is an obstacle to that argument, an obstacle that exists exactly at the point (as Chakrabarty has stressed) at which it becomes necessary to regard that planetary "we" as not only *impacted* by the geophysical force of climate change but *as* that self-impacting, self-reorganizing, geophysical force. By the terms of the argument I have been advancing, addressing that obstacle comes down to this question: when we regard the image of this young child, standing in the ruins of his ocean-risen, beach sand–devoured home, do we see a visual field that is ontologically single or a field that is ontologically split? Do we see a world composed of two forms of matter? Human matter and natural matter, arrayed in their long enmity toward one another, with natural matter—the sea—now triumphing over human matter—the boy? Or do we see a single common material plane? A world in which Collins Kusietey—regardless of his will in this planetary case and drama—has slipped, burst, vanished from the distinctive ontology of the human, as the sea has also lost its own "natural" ontic distinctiveness, and the two (in one "higher" level or scale of inspection) have merged in a hybrid humanatural fabric of geophysical force and ontic matter?

That is the question—vital and inescapable—Chakrabarty's argument forces.

And to resolve it, I need now to alter the core question I have borrowed and adapted from Sartre. The question can no longer simply be: "What, now, can we know of Collins Kusietey?" but also, instead and simultaneously, "What, now, can we know of the sea rising up behind him?" How are we to understand it? What matter of matter does it encompass? Which of the competing histories of materialism are adequate to account for its fabric of being?

Is it—to preview the discussion ahead—best understood, in Sartre's terms, as a counterfinality? Or, in Timothy Morton's, as a hyperobject?

Or, as ever, both at once?

In either case, what difference would that difference (or non-difference) make to our understanding of these menacing waters? What difference would it make to our understanding of the new and enduring and vanishing "we" of the human? What difference would it make (whether in space, in time, or in being) to the problems of justice and freedom that are evident in Collins Kusietey's so obviously unjust and so obviously unfree exposure to the sea from which he and his fellow citizens are no longer distant?

"There are no material objects which do not communicate among themselves through the mediation of men," Sartre insists in the *Critique of Dialectical Reason*:

> And there is no man who is not born into a world of humanized materialities and materialized institutions, and who does not see a general future prescribed for him at the heart of the movement of History. In this way society in its most concrete movement is shot through with passivity, and unceasingly totalizes its inert multiplicities and inscribes its totalization in inertia, while *the material object, whose unity is thereby recreated, rediscovered and imposed, becomes a strange and living being with its own custom and its own movement.* . . . *This vampire object constantly absorbs human action, lives on blood taken from man and finally lives in symbiosis with him.*[28]

These dense and surprising passages (dense in the inevitable ways in which Sartre's painstakingly developed arguments depend not only on their logical organization but on the particular meanings he ascribes to key terms in his personal theoretical lexicon—"passivity," the "inert," "totalization," etc.—and surprising in their seeming anticipation, fifty years before the fact, of a contemporary "new materialism's" vision of a world of animate, vibrant matter surrounding, infiltrating, and living in "actant" "symbiosis" with the world of "human action") are taken from what Frederic Jameson characterizes as one of the *Critique*'s most famous "set pieces." In it, as Jameson indicates, Sartre is attempting to explain his notion of a "praxis without an author," or as Sartre also says, defining what he means by a "counter-finality": one of those apparently inert background domains or environments of human activity (in this case the mountainous and riverine landscape of mid-twentieth-century rural China), which has been so transformed by *prior* human actions (here, centuries or even millennia of deforestation) that it has trespassed the boundary between material background and human foreground and taken on, in Sartre's terms, the character of a new "strange and living being."[29]

Deforested, Sartre insists, the Chinese landscape is neither dead nor inert. Nor is it without the capacity for action. Rather, in its very devastation, it has become practico-inert. It has become capable of a practice. Or, more precisely still, it gives every appearance of having become the nonhuman "author" of a "praxis"; the inhuman author of a "project" with a determinate orientation toward the future.

What is that future-fashioning praxis of China's deforested mountainsides and river-valleys?

Flooding. Systematic, recurrent, devastating flooding. Social-, cultural-, lifeworld-reorganizing flooding, as year after year the rains sweep down over the denuded mountainsides.

> [S]ince the loess of the mountains and peneplains is no longer retained by trees, it congests the rivers, raising them higher than the plains and bottling them up in the lower reaches, and forcing them to overflow their banks . . . Thus, the absence of trees, which is an inert and thus a material negation, also has the systematic character of a *praxis* at the heart of materiality.[30]

Thus what had been mere "matter," the substance of the "geological and hydrological structure of China," becomes a "humanized materiality," a "counter-finality" not only accumulating (absorbing) within itself the past histories of human action, but, on the "rebound" of accumulating those impacts within itself, turning their consequences against the present and future organizations and possibilities of human life.

Thus ecological catastrophe inscribes a malevolently living matter as a *prescribed* "general future" at "the heart of the movement of History."

And thus, the inevitable question arises. Climate change, the geological reorganization of the globe, the rising of the planet's seas: are all of these a massively scaled-up version of such a deforested/Sartrean counter-finality? Such a practico-inert, humanized materiality residing at the heart of the future movement of global history?

Or is climate change (manifest in the rising waters looming up on Collins Kusietey's shore) better understood, in Timothy Morton's terms, as a *hyperobject*? From Morton's perspective, the answer is clear, as is evident in one of his earliest definitions of the concept, where he includes as exemplary instances of what he has in mind, phenomena "such as radioactive materials and global warming . . . [which] stretch our ideas of time and space, since they far outlast most human time scales, or they're massively distributed in terrestrial space and so are unavailable to immediate experience."[31] As his subsequent and more expansive work on this question indicates, however, while he certainly regards global warming as a hyperobject (perhaps the hyperobject par excellence), to regard it as a phenomenon is to miss the point entirely. That is because to speak of global warming, the rising seas, or any other hyperobject as phenomenal is to remain within a "correlationist" epistemology whose authority Morton, and the broader project of Object Oriented Ontology (ooo) with which his work is affiliated, comprehensively rejects.

The problem with correlationism, as he argues—in line with that range of writers I am associating with the emergence of Materialism II (Bennett, Harman, Latour, and Meillassoux, among them)—is that it depends on "the notion that philosophy can only talk within a narrow bandwidth, restricted to the human-world *correlate*: meaning is only possible between a human mind and what it thinks, its 'objects,' flimsy and tenuous as they are."[32] To restrict philosophical discourse so, to limit it to the dyad of the "human" thinking (or phenomenologically being in) the "world," Morton further indicates, is to fundamentally misrecognize the reality that hyperobjects such as global warming demand we acknowledge: the end of the world is no longer an image of science fiction: "The end of the world has already occurred."[33] By that, Morton stresses, he does not mean that the earth has "exploded" but that the "world" as exterior object relative to the enclosure of the human (the world as outside-object in relationship to which the inside-object of the human exists) no longer stands—and has not stood since the dawning of the age of the Anthropocene.

> The end of the world has already occurred. We can be uncannily precise about the date on which the world ended. Convenience is not readily associated with historiography, nor indeed with geological time. But in this case, it is uncannily clear. It was April 1784, when James Watt patented the steam engine, an act that commenced the depositing of carbon in Earth's crust—namely, the inception of humanity as a geophysical force on a planetary scale.[34]

From that moment on, Morton insists—in terms fundamentally consistent with the discourse I have been sketching—"we" [we humans] are not apart from "the world" [out there]. We are more than situated in it, more than impacted by it, more than determined by it, more than correlated to it. We and it do not exist on separate planes. There has been, instead, "a quake in being": a quake that has rendered all things "interobjective"; a quake that has annulled the distinction between "foregrounds" and "backgrounds" on which the separatist correlation of the "world" and the "human" depends.[35]

> Now let's think the evaporation of *world* from the point of view of *foreground* and *background*.... What happens when global warming enters the scene? The background ceases to be background ... [and if] there is no background—no neutral, peripheral stage-set of weather ... then there is no foreground.... The specialness we granted ourselves as unravelers of cosmic meaning, exemplified in the uniqueness of Heideggerian Dasein, falls apart since there is no meaningfulness possible in a world without a foreground-background distinction. Worlds need horizons, and horizons need backgrounds, which need foregrounds.[36]

It is here—with this turn to the loss of the foreground-background distinction between the human and its others ushered in by climate change—that we begin to get a sense of both the similarity and the ultimate difference between counterfinality and hyperobject, and of the distinction and similarity between the alternate forms of matter their alternate materialisms contemplate.

For all the contamination of human history effected by Sartre's "humanized materialities"—for all the ways in which *apparently inert background environments have trespassed the boundary between material background and human foreground and taken on the character of a new "strange" and "living being,"* as I earlier put it—in Sartre's dialectic the boundary between the two realms (the realm of the world and the realm of the human; the realm of the background and the realm of the foreground) remains—even as it is trespassed. Indeed, that boundary is thrown into stark relief by being trespassed. Which is another way of saying that, at the end, Sartre's materialism, like the Marxist materialism on which it draws—the materialism I am calling Materialism I—remains correlationist (and in remaining so, reminds us that for all their profound differences, both Kantian idealism and classic historical materialism begin and abide with the foundationally correlationist and ontologically riven dyad of the human and the world).

This can be simplified.

For Materialism I, the background of the world may encroach into the realm of the human, but it does not annul the human. The sea may rise and push its vomit of sand into the quarters of human life. But the sea does not become interobjective with human life. *We* and *the sea* remain two distinct forms of matter, locked in fierce antagonism. The human remains, with its projects, its praxis, its determination to be free from the tyranny of the sea, free from the violence of that other form of matter.

For Materialism II, for Morton, for Object Oriented Ontology, that is a vain hope. There is no longer a human, no longer a "we" divisible "from" the sea: neither miles from the sea or millimeters away. There are no longer two modes of matter. Climate change, the ultimate and defining hyperobject, has annulled that difference. "We may scale up like this as far as we like. We will find that all entities whatsoever are interconnected in an interobjective system."[37] There has been a quake in being. There is no freeing ourselves from it. If freedom is to remain an object of thought and of action, it will need to orient itself toward this interconnection, toward what Morton names "the mesh," toward what Latour calls "the parliament of things."[38]

Indeed, it is exactly with regard to this notion of cosmological assemblage, as Mbembe details it in the final chapter of *Critique of Black Reason*, that we can see another clear articulation of the distinction between *Materialisms I and II* and of the differing stresses these two materialisms put on the problem of freedom: in this case—to return to one of my opening arguments—as exhibited by the differences between Mbembe's and Gilroy's conceptions of the demands placed on critical black thought by the advent of the Anthropocene and the dialectical shift Anthropocene thought demands from a primary discourse on the global to a merged discourse on the global *and* the planetary. For Gilroy, as I have indicated, the rising of the seas, the fortressing of Europe and the United States against climate-refugees (and subaltern migrants from all situations) demands the articulation of a new planetary humanism nourished by the traditions of Black Atlantic thought. In limning that planetary humanism, Gilroy takes inspiration from a range of thinkers, most notably Frantz Fanon, one of whose defining contributions, he suggests, was to pave the way for a "new humanism outside race," a "reparative humanism" forged in the midst of anticolonial struggle.[39] Key to that accomplishment, Gilroy argues, was that "[r]ather than simply holding humanism responsible for the development of racism, Fanon, who was hostile to what he saw as the fraud that followed from the reification of racial identity, approached racism instead as a major factor in the corruption of humanism."[40] Fanon's genius, Gilroy suggests, was to refuse the assumption that there was a constitutive "discursive tie" between humanism and racism, but, instead—like Edward Said decades later—to "focus on the abuses of humanism rather than its mechanistic refusal."[41] In drawing that conclusion, Gilroy places himself, and his call for a planetary humanism foundational to those "new collectivities and solidarities" requisite to our moment, squarely in this Fanonian (and Saidian) line.[42] "Even now," he concludes, "[Fanon's] arguments can renew our incentive to imagine a new humanism that has been contoured specifically by the denaturing of race and the repudiation of racial orders."[43]

Significantly, in making these arguments, Gilroy not only allies himself with a humanist Fanon but explicitly against recent currents of "posthumanism," typified, for him, by the work of Haraway. More specifically, he repudiates the tendency he sees in her (and other posthumanist) work to displace the "interspecies conflict[s]" of our moment (most virulently emblematized, in his Tanner Lectures and other subsequent work on this question, by the surgence and electoral-embrace, across an alarming swath of liberal-constitutional democracies, of a range of neo-right populisms and racialized anti-refugee policies) in favor of an "increased concern with intraspecies relations" (for which addressing the problematics of the human-animal [in]distinction and a post-correlationist ontology of life, survival, precariousness, and extinction becomes the primary task of critical thought).[44] "The influential work of Donna Haraway," he accordingly suggests, "is paradigmatic of the plea to employ interaction with our companion species as a mechanism to learn the transferable skills involved in 'living intersectionally.'"[45] While not dismissing (in fact, endorsing) such "commitment[s] to radical relationality and a political ecology that refuses the conceits of approaching nature as an exploitable, limitless resource," the post-humanist approach to such commitments, he suggests, falls error to the categorical error Fanon refused: the assumption, as Haraway writes, that: "'[t]he discursive tie between the colonized, the enslaved, the noncitizen, and the animal—all reduced to type, all Others to rational man, and all essential to his bright constitution—*is at the heart of racism and flourishes, lethally, in the entrails of humanism.*'"[46] The consequence of that line of reasoning, Gilroy indicates, is the flawed conclusion that if "[h]uman exceptionalism has underpinned the impending disaster that can be gauged in the looming catastrophe of the

Anthropocene or, more accurately, the Capitalocene," then any *humanist* response to that catastrophe (tainted by its inherent racism) must be abjured, while cross-species relationality must be privileged over a primary concern with the well-being, dignity, protection, and rights of the human.[47] And that is precisely the conclusion Gilroy refuses, and precisely why Fanon—who, for him, refused to accept that there is such a constitutive and unvoidable "tie" between humanism and racism—is so crucial to the planetary humanism for which he calls, the new humanism, which, via Gilroy's writing and advocacy, sits as perhaps *the* paradigmatic Black Atlantic Anthropocene contribution to the lineage of critical thought I am calling *Materialism I*.

For Mbembe, Fanon is also a central figure. But Mbembe's Fanon differs meaningfully from Gilroy's Fanon, as their materialist epistemologies also significantly differ. That said, there is a clear point of Fanonian convergence from which Mbembe's and Gilroy's subsequent differences depart. Whether *humanist* or not, Fanon is, for both, a spokesperson of "humanity" and of the struggle for liberty against subordination and abandonment with which any discourse on the human (whether in its humanist exceptionality or in its post-humanist entanglements) must contend. As Mbembe unambiguously states: "If there is one thing that will never die in Fanon, it is the project of the collective rise of humanity. In his eyes, this irreplaceable and implacable quest for liberty required the mobilization of all of life's reserves."[48] Where, then, do Mbembe and Gilroy differ? Where do their Fanon-inspired planetary materialisms diverge? The key is in the second of Mbembe's sentences, in his insistence that the Fanonian quest for liberty—in whatever moments, circumstances, or situations a freedom-questing "humanity" finds itself—does not begin with a reversion to (and repair of) the humanist tradition but with "the mobilization of all of life's reserves." By which, as he goes on to stipulate, he does mean all of "life's" reserves, human and nonhuman, in whatever determinate condition of exposed life humanity's most vulnerable and most "wretched" encounter those reserves of life coming to light (not as goods to be plundered but as resources, "in-common," to be cultivated and "shared").[49] In the colonial situation, those life reserves, Mbembe unflinchingly indicates, included, for Fanon, the reserve of anticolonial violence—though not, he insists, without Fanon's recognition that "by choosing 'counterviolence' the colonized were opening the door to a disastrous reciprocity—a 'recurring terror.'"[50] The key to closing that door, Mbembe indicates, was that for Fanon the purpose of violence was not violence itself, but that, instead, from his "perspective, the goal of struggle was to produce life. . . . It was, in effect, through violence that 'the *thing* which has been colonized becomes a man' and that new men could be created, along with 'a new language and a new humanity.' Life as a result took on the appearance of an unending struggle."[51]

Mbembe's object in advancing that reading is neither to extoll nor to dismiss the violence of anticolonial struggle for freedom against the "total war" violence of colonialism but to return our attention to Fanon's unrelenting commitment to the conviction that subaltern struggle will, and must, find its ultimate realization through the production of a "new humanity, through the production of "life" (and new "forms of life") in the most constrained of circumstances.[52] As he makes that point, Mbembe's purpose is also to remind us that Fanon must not be read nostalgically; that to read Fanon now is, indeed, to read him *now*, through *our* contemporaneity's challenges to freedom; that "to reread Fanon today. . . . [is only] partly about learning to resituate his life, work, and language within the history into which he was born and which he tried to transform through struggle and criticism. It also means translating—into the language of our time—the major questions that made him to stand up, uproot himself, and travel among companions along the new road that the colonized had to build with their own strength, with their own inventiveness, with their irreducible will. We must reactualize this marriage of struggle and criticism in our contemporary world."[53]

What does that contemporary reactualization of Fanon—and of the histories of black reason, struggle, and critical thought he represents—demand?

What does it demand from the argument of this book?

Not, for Mbembe, a reparation of humanism. Instead, it requires a radical commitment to travel among humanity's new companions, a radical resolve to mobilize (collectively, "in-common") *all* of life's reserves as a new (and renewed) practice of Fanonian reason rededicates itself in our planetary moment to the liberatory struggle not for an old but—once again—for a "new" humanity. What does that new humanity look like? No longer a humanity set against "the world" out there but, instead, as Mbembe indicates in his conclusion to *Critique of Black Reason*—in as powerful an articulation of an Anthropocene *Materialism II* and as strong an affirmation of Quarmyne's "*We Were Once* [but are no longer] Three Miles from the Sea" as I can imagine—a humanity that is simultaneously the humanity of the wretched of the *earth* and the humanity of wretched of the *eaarth*:

It is therefore humanity as a whole that gives the world its name. In conferring its name on the world, it delegates to it and receives from it confirmation of its own position, singular yet fragile, vulnerable and partial, at least in relation to the other forces of the universe—animals and vegetables, objects, molecules, divinities, techniques and raw materials, the earth trembling, volcanoes erupting, winds and storms, rising waters, the sun that explodes and burns. . . . There is therefore no world except by way of naming, delegation, mutuality, and reciprocity . . . And so the difference between the world of humans and the world of nonhumans is no longer an external one. In opposing itself to the world of nonhumans, humanity opposes itself. For, in the end, *it is in the relationship we maintain with the totality of the living world that the truth of who we are is made visible.*[54]

Or, as Mbembe allegorically explains the figure of "cosmological assembly" underpinning his account of this new materialism:

In ancient Africa the visible sign of the epiphany that is humanity was the seed that one placed in the soil. It dies, is reborn, and produces the tree, fruit, and life. . . . It was understood that nature was a force in and of itself. One could not mold, transform, or control nature when not in harmony with it. And this double labour of transformation and regeneration was part of a cosmological assembly whose function was to consolidate the relationships between humans and the other living beings with which they shared the world.

Sharing the world with other beings was the ultimate debt. And it was, above all, the key to the survival of both humans and nonhumans. In this system of exchange, reciprocity, and mutuality, humans and nonhumans were silt for one another.[55]

With that, we are close to closing the loop.

In considering the rising seas harrowing the coast of Ghana alternately as a counter-finality and as a hyperobject, I have sought to tease out the similarities and differences of two alternate materialist approaches to the "planetary conjuncture" of climate change. For the first, Materialism I, the sea (and, more generally, the climate change it manifests) constitutes a form of matter profoundly affected by human action, drawn into a devastating feedback loop with human action, but ultimately ontologically distinct from the human. With regard to the scales of time and being I have been discussing, Materialism I thus cleaves the Anthropocene order in two. On one side there is the biographical and the nomological co-determining one another; and on the other side there is the biological, the zoëlogical, and the geological, themselves altered by human biographies and the human orderings of the earth and, in their turn, altering the course of individual and collective human life, but not fully blending with the human scene. For Materialism II, conversely, there is nothing but the blend: the Anthropocene order is all interconnection—the biographical meshed with the geological, the nomological with the biological, the zoëlogical with the cosmological in a great weaving together of all these scales of time and being.

That is one way to compare and distinguish these two modes of apprehending our planetary condition and the matter of our times.

However, as the stress I have laid on the word *freedom* indicates, there is another way of distinguishing these two materialisms, and the invitations they make, in Marx's terms, not only to interpret the Eaarth but to change it—not only to apprehend our condition but to act within it and to make something emancipatory of it. That difference is the difference in their orientations to freedom.

In the broadest sense, the discourse of freedom proper to Materialism I remains within the ambit of the long-standing classical debate on negative and positive conceptions of liberty, with the proviso that in the time of climate change both of those concepts find themselves stressed by a new form of pressure. If, in other words, negative liberty is characterized above all else—in a discourse descending from Kant to Isaiah Berlin—as a freedom from arbitrary sovereign constraint, then to that lineage's defense of liberty as a mode of freedom safeguarded from the tyranny of the absolute state and absolutist religion, an Anthropocene valorization of such freedom adds as a regulatory and aspirational ideal a project of human being liberated from the arbitrary sovereignty over the planet of CO_2 counts and melting ice, rising seas and heat waves, advancing deserts and collapsing fisheries. This is the theory of liberty and human rights at the heart of the 2015 Paris Climate Accord: a multinational agreement predicated on a classically humanist materialism and a pre-posthumanist understanding of the actor and the situation.

Look again at Quarmyne's images of the village of Totope with that in mind. See those images as a visual manifesto of the Paris Accord's politics of negative liberty, as a documentary art's urgent reminder of the right of

Collins Kusietey—and all the inhabitants of his town, and all the citizens of a planetary-wide archipelago of postcolonial dwelling places—to be free from the violence of the waters crowding on the horizon, free from the sovereignty of ice, free from the "slow" and sudden "violence" of carbon.

Then see those images again, not only as emblems of negative liberty (not only as an exhortation to be free from a circle of arbitrary sovereign power that has grown, over the history of the modern, from the sovereignty of the church, to the sovereignty of the state, to the sovereignty of polar ice) but as a renewed call to positive liberty; as an exhortation from the twenty-first-century West African shore that we not abandon the constructive power of the social and political freedoms promised by the mid-twentieth-century anticolonial movements. This is the positive freedom partly realized in Ghana in 1957 as the country became the first independent African state; the freedom, as succeeding generations of Ghanaian and West African writers, artists, and political activists have maintained—and as Quarmyne's photographic series so clearly wishes to remind us—that remains incompletely realized and urgent to fulfill.

The Beautyful Ones Are Not Yet Born, declared Ayi Kwei Armah in the title of his 1968 novel of post-independence Ghana, encapsulating the frustrated promise of independence together with the hope that that generative promise might "yet" be realized.[56] Armah's central emblem of that unrealized hope of a new coming-into-being is "the man," the novel's protagonist, and his key image of postcolonial disappointment is the man's flight through a gap of sewage, a fecal morass metonymic of the state. Not fully an ecological novel, *The Beautyful Ones Are Not Yet Born* nevertheless opened the door to an ensuing body of West African work explicitly linking the unfulfilled projects of postcolonial liberty with environmental freedom. The categorical example of this tradition of work is the writing, life, and death of Ken Saro-Wiwa in his struggle for the right of the southeast Nigerian Ogoni people to a heightened degree of self-determination, compensation for the oil extracted from their lands by Royal Dutch Shell, and ecological de-devastation.[57] His dream of both a negative *and* a positive experience of postcolonial and environmental freedom (both a freedom from and a freedom to) is, in a meaningful sense, echoed and endorsed (two decades after Saro-Wiwa's execution) by the Paris agreement, with its acknowledgment that "climate change is a common concern of humankind," and the commitment, by all its signature parties, "when taking action to address climate change, [to] respect, promote and consider . . . human rights, the right to health, the rights of indigenous peoples, local communities, migrants, children, persons with disabilities and people in vulnerable situations[,] and the right to development."[58] Genealogically filiated to Morrison's *Beloved* and the diasporic tradition that work represents, "We Were Once Three Miles from the Sea" also belongs in this lineage from *The Beautyful Ones Are Not Yet Born*, to the Movement for the Survival of the Ogoni People, to the Paris agreement.

Which is another way of saying that kin to Morrison's Sethe and Denver and Baby Suggs, Collins Kusietey is further kin to Armah's "man": also one of the beautiful ones waiting to be born into the realization of positive freedom—freedom to health, freedom to shelter. . . .

We cannot regard him otherwise.

And yet: in the era of the Anthropocene, that positive "freedom to" is profoundly vexed.

For if, at the scale of the child, at the scale of the individual life, at the scale of biography, the right to that freedom seems indispensable, as we begin to scale up (to the scale of an entire society, the scale of the national and transnational, the scale of an entire global *nomos of the Eaarth*), that freedom—what the Paris document calls the right to "development"—and the modern system of fossil fuel–based energy on which "development" depends (on which hospitals depend, on which housing construction and electrified schooling depends) enters into a feedback loop with the forcings of climate change to exacerbate the conditions of unfreedom so evident in the ruined villages of Quarmyne's Ghanaian shore.

This is the tragic irony Chakrabarty had in mind when he wrote, with devastating understatement, "the Mansion of modern freedom stands on an ever-expanding base of fossil fuel. . . . [M]ost of our freedoms so far have been energy-intensive. . . . [W]hatever our socioeconomic and technological choices, whatever the rights we wish to celebrate as our freedom, we cannot afford to destabilize conditions (such as the temperature zone in which the planet exists) that work like boundary parameters of human existence."[59]

What then is to be done?

Recognize, as I suggested earlier, the tragic knowledge that modernity's great projects of freedom—even, ultra-ironically, the late-modern project of freedom articulated by the Paris climate agreement—are indissociable from the catastrophe leading to (and not out of) the Anthropocene End of History.

Yes.

Consider the need to articulate a new theory of freedom, an alternate model of emancipation for Anthropocene times.

Yes. (I will return to this below.)

But abandon entirely these older conceptions of freedom, negative and positive liberties and their bundle of rights, and the theory of person and circumstance, of the actor and the situation (the Materialism I) on which they depend?

No.

They cannot simply be abandoned. They are part of the "blend." That is the argument I have been advancing throughout. To abandon these liberties, to abjure these rights, to jettison this materialism, is to misrecognize the full multi-scaled complexity of our situation. It is to avoid the numbingly difficult work of weaving together the biographical and the nomological with the biological, the zoëlogical, the geological, and the cosmological. It is to pretend that the first two scales of being and time do not exist. It is to eschew a dialectics of forces and forcings in favor of the monopoly of forcing. It is to act as if being is without irony, without contradiction. And the virtue of holding to Materialism I—even as we seek, simultaneously, to stretch beyond it, even as we seek to hold it in knotted tension with Materialism II as Chakrabarty earlier held History I and History II in roped tension with one another—is precisely that this materialism refuses to free itself, or us, from contradiction. It does not resolve the tragic irony of our time. Its optic on the world illuminates tensions and paradoxes, and urges us to continue to do so, and to live with their difficulty.

What, in practice, does that mean?

Perhaps something as simple and complex as returning to another of the images in Quarmyne's haunting series. A mother and her three young children hold flickering cans of paraffin cupped in their hands, their visages triply illumined above the sewage-thick floodwater in which they are standing: illumined once by the flames in their hands, a second time by the take of Quarmyne's camera, and a third time by the wattage powering our energy-devouring laptops and computer screens as this image enters into global circulation and visibility at https://phmuseum.com/nyaniquarmyne/story/we-were-once-three-miles-from-the-sea-c09d2397e6.

The irony and the tension are not difficult to grasp. The condition of this family's illumination (on the screen) is the condition of a globally interlinked energy and information system. Without that system, the night of darkness

Figure 3.11. A Family in Azizanya. From "We Were Once Three Miles from the Sea." © Nyani Quarmyne/Panos Pictures.

closes around their plight, closes around their irrefutable and unrealized rights (to be free from that fecal water in which they are standing; to be free to realize every possibility of human flourishing each of their four lives so utterly demands). And yet, the system that illuminates their precarity is, of course, exactly the one that by rendering their situation globally visible, exacerbates it—as the energy pulses, and the wattage mounts, and the screens glow, and the waters rise.

Which does not mean that we are free to turn those screens off and look away from the eaarth they make visible and the eaarth they create. Nor does it mean that we are free to say that the right of a *freedom to* has ended; that it is restricted to those of us fortunate enough to be on this side of the glowing screen; that it can no longer be extended; that the right to shelter, and health, and schooling—and the energy on which all those rights, in practice, depend—cannot come to the Ghanaian shore, even as we know that as that dream of freedom comes shore to shore to all the distributed coastal towns, delta zones, megacities, and disregarded hinterlands of the global postcolony, so too does the wattage surge, the desert advance, the water rise....

None of this means that we can answer these paradoxes definitively. That is not the point of insisting that we continue to dwell with this orientation toward the world, this materialism, this paradoxical and unabandonable conception of rights and freedoms. The point of continuing to abide in this zone of tension is not to conclude an answer to these questions. It is to ask, to insist, that on whatever side of the screen and the paraffin lamp we live, we continue to hold ourselves in a condition of relation toward one another: across that line.

And then we need to take another step. We need to take from this invitation toward "entanglement" (in Sarah Nuttall's terms) a further invitation to enter the blend, or as Morton also has it, "*the mesh*."[60] As he indicates, this means not just a commitment toward that "intersubjectivity" implicit in the movement across the space of the screen and the lamp but toward an "interobjectivity" that the hyperobject of climate change renders visible, the interobjectivity that Materialism II renders visible.[61]

> [W]hat is called intersubjectivity—a shared space in which human meaning resonates—is a small region of a much larger interobjective configuration space. Hyperobjects disclose interobjectivity. The phenomenon we call intersubjectivity is just a local, anthropocentric instance of a much more widespread phenomenon. . . . In other words, "intersubjectivity is really human interobjectivity with lines drawn around it to exclude nonhumans."[62]

The message here—as Materialism I finds itself roped to Materialism II, as human intersubjectivity (individual and biographical, social and nomothetic) scales up to the domain of radical interobjectivity, trespassing the lines drawn around the human to mesh with the nonhuman (with bios and zoë, geos and cosmos)—is that the work of freedom does not end but, like time, accumulates: drawing more and more within its ambit, doubling, redoubling, and extending, toward . . .

Toward what?
Toward nonhuman-speciated life.
Toward life itself.
Toward the geological future of the planet.
Toward the cosmological horizon that Benjamin saw provincializing "the history of civilized mankind . . . [and] the stature which the history of mankind has in the universe."[63]
Toward "a new and universal solidarity" in connection "with all that exists," as Pope Francis put it in his encyclical, *Laudato Si'*.[64]

Toward the determination expressed in this papal letter—which I would include as the touchstone theological contribution to Materialism II—that "nothing in this world is indifferent to us."[65]

But if we are prepared to stipulate to this, one last question nevertheless remains.

Why is this a question of resolutely pursuing a *freedom toward* this massive "interobjective system" rather than a matter of simply *being with* all that exists? Why the political category and political imperative rather than an ontological claim and statement of fact? Why not just say we *are* toward? What constrains us from being so? Why do we need a political-ontology?

Because—to return to an earlier point—our current condition is one of unfreedom in our being toward the world in this way (in Morton's terms, a condition of unfreedom in our being toward the "afterness" of the world). A range of sovereign impositions and arbitrary powers stand actively in the way of this simultaneously humanist and radically post-humanist *freedom toward*.

What impositions?

The list is impossible to exhaust, but these stand out:

—the imperative of national security, with its ban against disregarding *our* security, regardless of who or where *we* are, and regardless of any will we may have to be *toward* and insecurely secure the good of our one-anothers;

—the biopolitical imperative underpinning the contemporary geopolitical order, which, as Judith Butler has examined, outlaws the precarity of the polis and forbids—by force-of-law, force-of-regulation, and force-of-arms—the sovereign polity's entanglement with and in and toward the precariousness of all (human and non-human) around it;[66]

—the imperatives of unrelenting, unending economic development, of the law of GDP (the "little big number," as Dirk Philipsen's history of Gross Domestic Product suggests, that has come to rule the world), and, relatedly, of what Pope Francis calls the "technocratic paradigm" based on the "lie" and the "false notion" that "an infinite quantity of energy and resources are available, that it is possible to renew them quickly, and that the negative effects of the exploitation of the natural order can be easily absorbed";[67] the imperative of the force-of-the-secular, which, many writers (Benjamin in his "Theses on the Philosophy of History"; Bennett in *Vibrant Matter*; Chakrabarty in *Provincializing Europe*; Francis in *Laudato Si'*; Latour in work ranging from *We Have Never Been Modern* to the Gifford Lectures; Lévi-Strauss in

The Savage Mind; Morton in *Hyperobjects* and *The Ecological Thought*) quite differently but surprisingly similarly conclude, places a ban on living in non-human time, sets beyond political recognition and consideration the intermingling of progressive time and the multi-scaled times of now-being; the time of actors and the time of actants, the time of the human and the time of the post-human, the time of history and the time of the gods, the time of the state and the time of the cosmos, the time of the biographical and the time of the geological, the time of forces and the time of forcings, the time of the Atlantic trade and the time of the Atlantic Ocean.

To return to the question: recognizing that we are entangled, recognizing that we are toward, recognizing that things are this way, recognizing that we are enmeshed in a vast interobjective system made visible by the hyperobjectivity of climate change is not enough.

We need to pursue the condition to be *free* to be so; to be, in Butler's terms, precariously undone toward all that exists in this way.

And what this implies is that the urgent *freedom toward* of the Anthropocene (the new conception of freedom implicit in the Materialism II of our age) needs to take and blend into itself elements of the two freedoms associated with the older Materialism I: freedom from arbitrary sovereign command alongside with and braided through the freedom to realize the possibilities of collective well-being, planet-wide, across the conditions of our mutual precarity as the carbon accumulates, and the temperatures mount, and the oceans rise toward the condition of History 4°C.

Or, to return one last time to the figure of Collins Kusietey, standing on his shore: What now, from the vantage of the discourse of freedom do we see?

The imperative to be free *from*

The imperative to be free *to*

And the imperative *addressed to all of us—whoever "we" are, wherever "we" find ourselves*, however far "we" are in time or place from this coast from which we are no longer distant, but which the time of the Anthropocene has moved so near to us that we have no separation from it, or its child, or its sand, or its slave factories, or its surging water—to *be* toward, *act* toward, *live* toward what we have experienced ourselves, a quake in our being together, to be.

Coda. The Youngest Day

"The end of all things," Immanuel Kant suggested in a 1794 essay by that title, will come in one of three guises. It will come as one of three "last days" jostling and competing with one another for sovereignty over the course and end of human history: a "supernatural" last day, a "perverse" last day, or a "natural" last day. It is the task of philosophy, he further argues, to ensure that one of these days will triumph over the others, to ensure the final historical victory of the "natural" last day.

By "natural last day" Kant does not intend the day of nature. As might be predicted from the inclusion of "The End of All Things" among the far more famous set of essays comprising Perpetual Peace ("Idea for a Universal History with a Cosmopolitan Intent"; "On the Proverb: That May Be True in Theory, But Is of No Practical Use"; "An Answer to the Question: What is Enlightenment?"; and the eponymous title piece), the natural last day—which is also the faculty of philosophy's victory day—is, in fact, the day of reason's triumph over what is physically "raw" and "animal" in humanity. It is the day of the cosmopolitan, the day of enlightenment, the day (as he notes in "Speculative Beginning of Human History," another of Perpetual Peace's less discussed essays) in which "mankind" completes "the transition from the raw state of a merely animal creature to humanity, from the harness of the instincts to the guidance of reason—in a word, from the guardianship of nature to the state of freedom."[1] The "natural last day," as Kant clarifies in "The End of All Things," might thus better be identified as what "our language has chosen to call . . . the youngest day [*jungster Tag*]": "the point in time that concludes all time" as humanity, through the guidance of reason, and under the tutelage of the philosophy faculty, "end[s] the conflict within [itself] as [a member of both a] moral species and a natural species" and takes on that "second nature, which is the final goal of the human species' moral vocation."[2]

Or potentially so. For as it is the fundamental purpose of "The End of All Things" to suggest, humanity's realization of itself (and its moral species-being) in the passage from captivity-by-nature to a universally cosmopolitan "state of freedom" is by no means assured. The youngest day, the day of enlightenment, remains a day whose coming is not yet secure. It remains but one of three possible endpoints toward which humanity is headed over the long course of its emergence from a state of nature. And as might be expected from the ways in which "Speculative Beginnings of Human History" and "The End of All Things" cast the epistemological conflict between the medical and philosophy faculties (the struggle that Kant was to outline in the 1798 volume, *The Conflict of the Faculties*) as a narrative of humanity's long struggle between animal life and the capacity for reason, the other two "last days" in historic contention with the youngest day find their epistemological ground in the two additional "higher" faculties with which he held the philosophy faculty to be in eternal conflict: theology and law.

The youngest day, in other words, represents philosophy's day of triumph, while the supernatural last day, Kant avers, is the end of history toward which theology is steering mankind, and the perverse last day is the legal faculty's end of all things.

Under what signs will those other two last days appear?

The supernatural last day, Kant argues, is the day of the reign of miracles. It is the day of the blowing of angel trumpets, the day Saint John of Patmos described in his revelation. It is, by its other name, the "day of eternity," the day "with which [the light of reason] goes out and all thinking ends," the day in which, as a definitively "thinking subject," the human subject will find itself at once petrified and annihilated: "for then, surely, nature in its entirety will be fixed and, as it were, petrified; the last thought, the last feeling will come to a standstill in the thinking subject and remain, without change, always the same."[3] If humanity's "life" in the state of nature might be described, on Kant's terms, as animally pre-human, then, by his account, theology would direct history toward a last day in which humanity becomes something post-thinking, post-reasoning, post-human.

And the "perverse" last day? It will come, Kant warns, when the "love" that even philosophy can show toward the "moral constitution" of Christianity (for its capacity to demonstrate what we "ought to do" through "the free integration of the will of another into one's maxims") becomes impossible as, abandoning its "gentle spirit," Christianity takes on state-institutional form, acquires the power of command, and arms itself with "dictatorial authority."[4] If the private name of philosophy's last day is the "youngest day," and the supernatural last day is, alternately, the "day of eternity," then the perverse last day, Kant argues, is the day of the tyrant, and alarmingly so. For while tyranny may run perverse to the moral telos of humanity, it finds itself equipped to do so "armed" by a "code of laws" not based on reason but solely "on the command of an external legislator," a code and a power of command the faculty of law has the obligation to teach and the duty to obey but not the capacity to test for conformity "with reason."[5]

Hence, for Kant, the urgency of philosophy's desperate, lonely, and unimpeachable struggle for human freedom. Hence the clear sense that emerges in the conversation between this and his other essays on the faculties of the university and his contemporaneous writings on the philosophy of history over the last decade of the eighteenth century that the "conflict of the faculties" amounts to far more than a commentary on a set of internecine academic disputes. It highlights instead an epochal fight for the future, a fight for freedom that Enlightenment philosophy finds itself waging on three simultaneous fronts: against humanity's raw animality and captivity by nature; against the intellectual petrification of humanity in the becoming-choir of religious adoration; against the coming of the tyrant.

So, then: four faculties; one day of humanity's "animal" beginning; and three competing "last" days toward which the species is heading. Collectively, these elements ground Kant's complex allegory of the modern work of philosophy, philosophy's emergence in contradistinction to humanity's "natural species-being," and the ethical fight for the future of Enlightenment, which philosophy obliges itself to pitch against theology and the law, against church and dictator. Thus Kant writes the story of modern knowledge, of Enlightenment's struggle to bend the arc of history toward humanity's ultimate (cosmopolitan and republican) rendezvous with liberty.

But that story no longer holds. Not only because it is impossible to retain its starting point—no longer credible to assume the discrete existence and ontological difference of that raw, animal domain of "Nature" from whose "harness" philosophy offers to free reasoning humanity, and in opposition to which philosophical critique defines its domain of reason and freedom—but also because to Kant's three last days, our contemporary perceptions of the "end of all things" have added a fourth. Not the day of the republic, or the day of eternity, or the day of the tyrant, but the day of the Anthropocene: the "day" ironically dateable almost precisely to the moment in which Kant was writing; the day, by many accounts, beginning with the 1784 invention (seemingly beneath philosophy's level of attention) of the steam engine, the coincident wide-scale modern adoption of fossil fuels, and the launch of the industrial age; the day of the definitive merging of those domains of "natural" and "human" history on whose distinction Kant's entire enterprise (and much of the architecture of "modern" knowledge) depends; the day in which humanity, far from having exchanged its "natural" species-being for a "moral" species being, begins to so profoundly impact the climate of the earth and the geological future of the planet that, in Chakrabarty's terms, it comes to slip its ontological boundaries and become a globally destructive, incipiently catastrophic "force of nature."[6]

Against the coming of this last day of the species and this potentially catastrophic last day of the planet, what does philosophy have to say? Confronting the looming sovereignty neither of the dictator nor of petrified religion but of the climate, what is the task of critique? If philosophy has previously aspired to free humanity from animality, tyranny, and divinity, can it now affect the planet's temperature? Can the philosophy faculty's great multi-century project of freedom not merely survive the crisis of climate change, but so vigorously survive as to make of philosophy a countervailing, planetary force of nature?

It can. Though only, I have been intimating, if we are capable of grasping both our continuities and discontinuities with Kant's moment of writing. Only if we are capable of assessing the multi-scaled temporal, epistemological, and ontological complexity of a "situation" the Königsberg philosopher may not have realized he inhabited but which stretches, nevertheless, from his time into our own (and, potentially, for centuries or millennia to come). Only if, in assessing that situation, we hold to Kant's crucial insistence on linking philosophy to the struggle for freedom while simultaneously resolving philosophy's conflicts with its erstwhile antagonists, acknowledging its inescapable entanglements with the inseparable realms of "nature," law, and theology. Only if, set into concert rather than conflict, these faculties can fashion a new project of freedom adequate to the multi-scale complexity of the situation of the Anthropocene: a project oriented toward a radical rethinking of freedom at the biological, geological, and cosmological boundary zones of human existence.

But if these faculties of knowledge—our faculties of knowledge, that is, we ourselves—are to do so, I want to suggest in conclusion that it may be worth returning, however counterintuitively, not so much to Kantian philosophy as to a worldly project for philosophy that his thought surfaces: a project implicit in such phrases as "the task of philosophy," "the contribution of philosophy," etc. It will already be apparent that while I neither embrace the conclusions Kant draws from his allegory of philosophy in "The End of All Things" nor his starting point for that fable (assuming as it does a world firmly divided between "nature" and its "political" opposite), I nevertheless find the rhetorical structure of that "Last Days" allegory useful. It will also be apparent that I have been speaking of "philosophy" in at least two senses that are broadly consistent with Kant's use of the term in *The Conflict of the Faculties*.

First, it signifies the disciplines of knowledge gathered together, in his model for the university, as the "philosophy faculty." In this sense, "philosophy" includes not only the contemporary academic discipline bearing that name but all the disciplines of the modern liberal arts whose method is fundamentally interpretive or hermeneutic. Used in this way, when I speak of the task of philosophy I am speaking of the task of the humanities, the interpretive social sciences, and the arts in understanding and addressing the challenges of the Anthropocene.

My second form of usage aligns philosophy less with a collection of disciplines (or an interpretive faculty) than with a particular habit of critique. In this I take inspiration, among other sources, from Judith Butler's reflections on the continuing significance of *The Conflict of the Faculties* to our distinctly post-idealist critical moment. Can "a passage through Kant" and through his "sense of the meaning of critique," she asks in "Critique, Dissent, Disciplinarity," "lead to political and social consequence" without first obliging us to "[subscribe] to a transcendental form of argumentation"?[7] If it can, she answers, that will be because in defining philosophy in opposition to the tyrannical power of the state, "freedom from state constraint comes [for Kant] to define the disciplinary task of philosophy . . . [to serve] as a constitutive precondition of philosophy's claim to free and open inquiry."[8] As philosophy thus comes not simply to benefit from but to name "the moment when state power retracts—or should retract—its commands or submits its own commands to a certain testing by a form of reasoning that is not itself furnished by the state," and, therefore, to name "the moment in which reason, defined as the power to judge autonomously, establishes the possibility for political dissent," we can, Butler concludes, "see at least two points [in Kant's position] that are worth underscoring for the purposes of negotiating the present":

> First, the operation of critique takes place within the discipline of philosophy, but it also takes place every place and any place its distinguishing questions get posed, so critique belongs not just to the discipline of philosophy but, as Derrida has insisted, throughout the university. Second, the operation of critique takes place not only in the identifiable domains of philosophy and within the walls of the university but every time and any time the question of what constitutes a legitimate government command or policy is raised.[9]

In this way, as "philosophy" comes to name an entirely immanent form of critique, it becomes, simultaneously, the task not only of a discipline, or of a faculty of disciplines within the university, but the task of the university itself *and* the task of a "civil community" now working in concert with the "learned community" of the university.[10]

When I speak of "the task of philosophy" that is also very much what I have in mind, with two important points of emphasis. First, that while the tyrannical state must necessarily remain an object of philosophical critique, the state can no longer be regarded as defining the outer limit of critique, as monopolizing philosophy's condition and circumstances of articulation. Rather, in the epoch of the Anthropocene, climate change (and the conditions of unfreedom and violence it produces and exacerbates) joins tyranny in demanding philosophy's renewed coming-into-being, in demanding a labor of thought as attuned to the geological condition of the planet as to the condition of the law. Second (to reiterate and expand the point that Butler takes from Derrida), for that philosophical work to acquire the force it needs, it can no longer be understood as the special privilege of one faculty of knowledge set in conflict with others. It must instead be a philosophical labor emerging from the concert of the faculties and the further concert of the civil and the learned communities.

If, in Butler's terms, I thus wish to "make use of Kant against Kant" (as she says we "can and must" do), then, as will now be quite apparent, I have also wished, throughout this work, to make use of Sartre against Sartre in drawing from him a third and last source for what I intend when I am speaking of philosophy. When I began work on this text I had no idea of the importance Sartre's writing (particularly the *Critique of Dialectical Reason* and its introductory *Search for a Method*) would come to assume in what I was trying to understand and to say. It was my immense good fortune, as I was beginning to formulate my thoughts, to have a conversation with Ainehi Edoro, who suggested that what I was trying to frame in terms of the play of the biological, nomological, geological, and cosmological scales of the Anthropocene had echoes in Lévi-Strauss's critique of Sartre in *The Savage Mind*. It had been some time since I had thought with any care about either Sartre or Lévi-Strauss, and neither have been much on the general horizon of critical theory for some time. Turned back, at Edoro's invaluable suggestion, I have found in that debate not only an avenue into the dialectic of freedom and circumstance but two additional things. Initially, and almost immediately, the recognition that if I lacked answers for the questions I wished to pose, it was, fairly simply, because I had no clear understanding of the method I wished to pursue in answering them. What I was struggling with, it became clear, was a search for a method adequate to a new world of circumstance, a new "situation," a new "Eaarth": a search for a method capable of responding to the question of what "philosophy" might now have to contribute to the question of what it means to make history, to fashion a future, in circumstances inherited from a near, a middle, and a deep past and from an equally deep future to come. As much as I have disagreed with Sartre throughout this volume, I have taken this foundational insight and framework from him.

Furthermore, I have taken from his *Search for a Method* a third and final definition of philosophy, an understanding that philosophy entails more than a discipline or set of disciplines and more than a habit of critical thought. It entails also, in his rephrasing of Marx's famous lines from the "Theses on Feuerbach" ("the task is not to interpret the world but to change it"), the understanding that philosophy encompasses, that philosophy is and must be, "philosophy-becoming-the-world."[11] The welcome news, here, of course, is that philosophy has a history of succeeding in that task. The day of tyranny remains the everyday of far too many people in far too many parts of the globe—as, worldwide, it remains a day always too much in danger of returning. But the day of the tyrant has also often been closed, or exploded, or held back. And philosophy, in all the senses in which I am using the term, has contributed no little part to that struggle against the tyrant's perverse day. If, however, in the day of the Anthropocene, philosophy is to continue to play that world-changing role, then it will now need to do so by extending its sense of what that world entails; by expanding, beyond idealist, or humanist, or standardly historically materialist boundaries, its grasp of what the world's "matter," "forces," and "forcings" entail. Or, to put it another way, if philosophy is to continue to be "philosophy-becoming-the-world" (as it can and must be), then it must also now teach itself, coincidentally, to be philosophy-becoming-the-planet, philosophy-becoming-the-climate, philosophy-becoming-the-world-at-4°C.

NOTES

CHAPTER 1. **Of Forces and Forcings**

1. See Stephanie E. Smallwood, *Saltwater Slavery: A Middle Passage from America to American Diaspora* (Cambridge, MA: Harvard University Press, 2008); Saidiya Hartman, *Lose Your Mother: A Journey along the Atlantic Slave Route* (New York: Farrar, Straus and Giroux, 2008); Bayo Holsey, *Routes of Remembrance: Refashioning the Slave Trade in Ghana* (Chicago: University of Chicago Press, 2008).

2. Smallwood, *Saltwater Slavery*, 35.

3. Paul Gilroy, *The Black Atlantic: Modernity and Double Consciousness* (Cambridge, MA: Harvard University Press, 1993), 221.

4. See Smallwood, *Saltwater Slavery*, 9–32.

5. Giovanni Arrighi, *The Long Twentieth Century: Money, Power and the Origins of Our Times* (New York: Verso, 1994); Hartman, *Lose Your Mother*.

6. Ian Baucom, *Specters of the Atlantic: Finance Capital, Slavery, and the Philosophy of History* (Durham, NC: Duke University Press, 2005).

7. Giorgio Agamben, *Homo Sacer: Sovereign Power and Bare Life*, trans. Daniel Heller-Roazen (Stanford, CA: Stanford University Press, 1998).

8. Justin Gillis, "Heat-Trapping Gas Passes Milestone, Raising Fears," *New York Times*, May 10, 2013, accessed October 14, 2013, http://www.nytimes.com/2013/05/11/science/earth/carbon-dioxide-level-passes-long-feared-milestone.html.

9. Gillis, "Heat-Trapping Gas."

10. Fredric Jameson, *A Singular Modernity: Essay on the Ontology of the Present* (New York: Verso, 2002), 29.

11. See Rob Nixon, *Slow Violence and the Environmentalism of the Poor* (Cambridge, MA: Harvard University Press, 2013); Ramachandra Guha, *Environmentalism: A Global History* (London: Pearson, 1999); *The Unquiet Woods: Ecological Change and Peasant Resistance in the Himalaya* (Berkeley: University of California Press, 1990); Elizabeth A. Povinelli, *Geontologies: A Requiem to Late Liberalism* (Durham, NC: Duke University Press, 2016); Dipesh Chakrabarty, "The Climate of History: Four Theses," *Critical Inquiry* 35, no. 2 (winter 2009): 197–222; "Postcolonial Studies and the Challenge of Climate Change," *New Literary History* 43, no. 1 (winter 2012): 1–18; "Climate and Capital:

On Conjoined Histories," *Critical Inquiry* 41, no. 1 (autumn 2014): 1–23; "The Planet: An Emergent Humanist Category, *Critical Inquiry* 46, no. 1 (autumn 2019): 1–31.

12. Perhaps the most significant of those disagreements comes from my sense that while climate change does indeed call us to rethink the question of the human under the category of "species," the profoundly uneven conditions of human vulnerability to the catastrophic effects of global warming also demand that we apprehend climate change as accentuating and intensifying long-standing patterns of human division between worlds of relative security and worlds of extreme precariousness. The time of the Anthropocene, I have thus come to think, is not so much a unitary time (of humanity's catastrophic species-being) as a mixed time in which individual, historical, biological, and geological orders of human (and nonhuman) being and time have entered into highly complex forms of relation with one another. See Chakrabarty, "The Climate of History: Four Theses."

13. Chakrabarty, "The Climate of History: Four Theses," 198.

14. I address the meaning and relation of these two terms, *force* and *forcing*, in detail in the sections that follow. Most broadly, by *force* I am referring to the powers of social, cultural, and political organization (and disruption) proper to the legal, economic, bureaucratic, and other institutions of "human" history; by *forcing* I have in mind the radiative pressures (from carbon dioxide emissions, sulfur, solar flaring, etc.) effecting changes in the mean surface temperature of the earth.

15. As I currently project it, that sequence of volumes involves three unfolding areas of focus: capital (in chapter 1), climate (in chapter 2), and war (in chapter 3).

16. See Friedrich Engels, "Ludwig Feuerbach and the End of Classical German Philosophy . . . With Notes on Feuerbach by Karl Marx 1845" (Berlin: Verlag von J. H. W. Dietz, 1888), 69–72.

17. See P. J. Crutzen, "Geology of Mankind," *Nature* 415 (2002): 23.

18. Jan Zalasiewicz, Mark Williams, Alan Haywood, and Michael Ellis, "Introduction: The Anthropocene: A New Epoch of Geological Time?," in *Philosophical Transactions of the Royal Society* (2011): 369, 838.

19. Zalasiewicz et al., "Introduction," 837 (emphasis added).

20. Zalasiewicz et al., "Introduction," 840.

21. David Archer, *The Long Thaw: How Humans Are Changing the Next 100,000 Years of Earth's Climate*, Kindle edition (Princeton, NJ: Princeton University Press, 2008), Kindle locations 551–53. I am indebted to Dipesh Chakrabarty's work for bringing Archer's scholarship to my attention.

22. Archer, *The Long Thaw*, Kindle location 555.

23. Archer, *The Long Thaw*, 40.

24. Archer, *The Long Thaw*, 76.

25. Archer, *The Long Thaw*, 6–7.

26. Archer, *The Long Thaw*, 157.

27. Subcommission on Quaternary Stratigraphy, "Working Group on the 'Anthropocene,'" Working Group Convenor: Jan Zalasiewicz, accessed December 9, 2019, http://quaternary.stratigraphy.org/workinggroups/Anthropocene/.

28. Subcommission on Quaternary Stratigraphy, "Working Group on the 'Anthropocene.'"

29. See Christina Sharpe, *In the Wake: On Blackness and Being* (Durham, NC: Duke University Press, 2016), 118 and throughout.

30. Sharpe, *In the Wake*, 106.

31. See Jane Bennett, *Vibrant Matter: A Political Ecology of Things* (Durham, NC: Duke University Press, 2010); Donna Haraway, *The Companion Species Manifesto* (Chicago: Prickly Paradigm Press, 2003); Bruno Latour, *Reassembling the Social: An Introduction to Actor-Network-Theory* (Oxford: Oxford University Press, 2009); Achille Mbembe, *Critique of Black Reason*, trans. Laurent Dubois (Durham, NC: Duke University Press, 2017), 181; Tim Morton, *The Ecological Thought* (Cambridge, MA: Harvard University Press, 2012), 15; Tim Morton, *Hyperobjects: Philosophy and Ecology after the End of the World* (Minneapolis: University of Minnesota Press, 2013); Pope Francis, *Laudato Si': On Care for Our Common Home* [Encyclical Letter] (Vatican City, Italy: Libreria Editrice Vaticana, 2015).

32. Chakrabarty, "The Climate of History," 199.

33. Dipesh Chakrabarty, *Provincializing Europe: Postcolonial Thought and Historical Difference* (Princeton, NJ: Princeton University Press, 2007), 6.

34. Chakrabarty, "The Climate of History," 208.

35. See Rob Nixon, *Slow Violence*. For reasons that will become evident in what follows, I have substituted Bill McKibben's neologism "Eaarth" for the standard denomination of the planet in Frantz Fanon's *Wretched of the Earth*. See Bill McKibben, *Eaarth: Making Life on a Tough New Planet* (New York: Henry Holt, 2010).

36. Karl Marx, *The Eighteenth Brumaire of Louis Bonaparte: Cambridge Texts in the History of Political Thought* (Cambridge: Cambridge University Press, 1996).

37. Ato Quayson, "The Sighs of History: Postcolonial Debris and the Question of (Literary) History," in *New Literary History* 43, no. 2 (spring 2012): 368–69. Quayson's essay is one of a series in a special edition of *New Literary History* responding to Chakrabarty's "The Climate of History" and Robert Young's "Postcolonial Remains," both published in the previous edition, *New Literary History* 43, no. 1 (winter 2012): 19–42.

38. IPCC, *Climate Change 2013: The Physical Science Basis*. Contribution of Working Group I to the Fifth Assessment Report of the Intergovernmental Panel on Climate Change, ed. T. F. Stocker, D. Qin, G.-K. Plattner, M. Tignor, S. K. Allen, J. Boschung, A. Nauels, Y. Xia, V. Bex and P. M. Midgley, eds. (Cambridge: Cambridge University Press), released September 27, 2013, http://www.climatechange2013.org/images/uploads/WGIAR5-SPM_Approved27Sep2013.pdf, accessed October 16, 2013.

39. IPCC, *Climate Change 2013: The Physical Science Basis*, accessed December 17, 2013.

40. Detlef van Vuuren et al., "The Representative Concentration Pathways: An Overview," in *Climatic Change* 109 (2011): 5–31, 20.

41. IPCC, *Climate Change 2013: The Physical Science Basis*, 15.

42. IPCC, *Climate Change 2013: The Physical Science Basis*, 14, 16, 16, 17, 17, 18, 19.

43. *Turn Down the Heat: Why a 4°C Warmer World Must Be Avoided*, A Report for the World Bank by the Potsdam Institute for Climate Impact Research and Climate Analytics (November 2012), http://climatechange.worldbank.org/sites/default/files/Turn_Down_the_heat_Why_a_4_degree_centrigrade_warmer_world_must_be_avoided.pdf), accessed October 14, 2013.

44. *Turn Down the Heat*, 1.
45. *Turn Down the Heat*, xv.
46. *Turn Down the Heat*, xv.
47. *Turn Down the Heat*, 15, 26, 62.
48. *Turn Down the Heat*, xv.
49. *Turn Down the Heat*, xvi, xv.
50. *Turn Down the Heat*, xvii.
51. *Turn Down the Heat*, xvii.
52. *Turn Down the Heat*, xv.
53. *Turn Down the Heat*, 60.
54. Giorgio Agamben, *The Open: Man and Animal* (Stanford, CA: Stanford University Press, 2004).
55. Slavoj Žižek, *Living in the End Times* (New York: Verso, 2011).
56. McKibben, *Eaarth: Making a Life on a Tough New Planet*.
57. Chakrabarty, "The Climate of History," 208.
58. Mbembe, *Critique of Black Reason*, 1 (emphasis added).
59. Mbembe, *Critique of Black Reason*, 6.
60. Mbembe, *Critique of Black Reason*, 2.
61. Mbembe, *Critique of Black Reason*, 2.
62. Mbembe, *Critique of Black Reason*, 3.
63. Mbembe, *Critique of Black Reason*, 3.
64. For Hegel's infamous dismissal of Africa from the philosophy of history, see G. W. F. Hegel, *The Philosophy of History* (New York: Dover, 2004).
65. Mbembe, *Critique of Black Reason*, 7.
66. Mbembe, *Critique of Black Reason*, 5.
67. Mbembe, *Critique of Black Reason*, 5–6.
68. See Jason W. Moore, ed., *Anthropocene or Capitalocene?: Nature, History, and the Crisis of Capitalism* (Oakland, CA: Kairos Press, 2016), and Sharpe, *In the Wake*, 106.
69. Mbembe, *Critique of Black Reason*, 6.
70. Paul Gilroy, "Suffering and Infrahumanity," Tanner Lectures on Human Values (2014), 28.
71. Gilroy, "Suffering and Infrahumanity," 21.
72. Gilroy, "Suffering and Infrahumanity," 21.
73. Gilroy, "Suffering and Infrahumanity," 28, emphasis added.
74. Achille Mbembe, "Decolonizing Knowledge and the Question of the Archive," https://wiser.wits.ac.za/system/files/Achille%20Mbembe%20-%20Decolonizing%20Knowledge%20and%20the%20Question%20of%20the%20Archive.pdf, accessed January 3, 2019, n.p.
75. Mbembe, "Decolonizing Knowledge and the Question of the Archive."

76. Mbembe, *Critique of Black Reason*, 4.
77. Gilroy, "Suffering and Infrahumanity," 28.
78. Mbembe, *Critique of Black Reason*, 6–7.
79. Jean Paul Sartre, *Search for a Method*, trans. Hazel E. Barnes (New York: Vintage, 1968), 91.

CHAPTER 2. **History 4° Celsius**

Epigraphs: Dipesh Chakrabarty, "The Climate of History: Four Theses," *Critical Inquiry* 35, no. 2 (winter 2009): 206; Timothy Morton, *Hyperobjects: Philosophy and Ecology after the End of the World* (Minneapolis: University of Minnesota Press, 2013), 5.

1. Claude Lévi-Strauss, *The Savage Mind* (Chicago: University of Chicago Press, 1962), 258–59.
2. Lévi-Strauss, *The Savage Mind*, 259.
3. Lévi-Strauss, *The Savage Mind*, 259.
4. Lévi-Strauss, *The Savage Mind*, 261.
5. Lévi-Strauss, *The Savage Mind*, 260.
6. Lévi-Strauss, *The Savage Mind*, 260.
7. Karl Marx, *Later Political Writings*, trans. and ed. Terrell Carver (Cambridge: Cambridge University Press, 1996), 32.
8. Jean Paul Sartre, *Search for a Method*, trans. Hazel E. Barnes (New York: Vintage, 1968), 45, 91.
9. Sartre, *Search for a Method*, 48, 53; David Sherman, *Sartre and Adorno: The Dialectics of Subjectivity* (New York, SUNY Press, 2008), 261.
10. Sherman, *Sartre and Adorno*, 261.
11. Lévi-Strauss, *The Savage Mind*, 260.
12. It is vital to note that, in this context, *epoch* does not refer to a geological epoch but to an extended period of human historical time. The relation of these two conceptions of *epoch* to one another emerges subsequently—and as a crucial concern of this book—as the geological turn to the discourses of the Anthropocene emerge over recent years.
13. Lévi-Strauss, *The Savage Mind*, 262.
14. Lévi-Strauss, *The Savage Mind*, 262.
15. Lévi-Strauss, *The Savage Mind*, 262.
16. Lévi-Strauss, *The Savage Mind*, 248.
17. Lévi-Strauss, *The Savage Mind*, 245, 246, 247, 245.
18. See James Chandler, *England in 1819: The Politics of Literary Culture and the Case of Romantic Historicism* (Chicago: University of Chicago Press, 1999), particularly the introduction and chapter 1.
19. Catherine Malabou, *What Should We Do with Our Brain?*, trans. Sebastian Rand (New York: Fordham University Press, 2008), 5 and throughout. For additional influential recent work in the neurohumanites, see also William Connolly, *Neuropolitics:*

Thinking, Culture, Speed (Minneapolis: University of Minnesota Press, 2002); Antonio Damasio, *Looking for Spinoza: Joy, Sorrow, and the Feeling Brain* (New York: Harcourt, 2003); Catherine Malabou, *The Ontology of the Accident: An Essay on Destructive Plasticity*, trans. Carolyn Shread (Cambridge: Polity Press, 2012); Catherine Malabou and Adrian Johnston, eds., *Self and Emotional Life: Philosophy, Psychoanalysis, and Neuroscience* (New York: Columbia University Press, 2013); Nikolas Rose and Joelle M. Abi-Rached, *Neuro: The New Brain Sciences and the Management of the Mind* (Princeton, NJ: Princeton University Press, 2013); Daniel Lord Smail, *On Deep History and the Brain* (Berkeley: University of California Press, 2007): Lisa Zunshine, *Why We Read Fiction: Theory of Mind and the Novel* (Columbus: Ohio State University Press, 2006); Lisa Zunshine, ed., *Introduction to Cognitive Cultural Studies* (Baltimore: Johns Hopkins University Press, 2010).

20. See Jane Bennett, *Vibrant Matter: A Political Ecology of Things* (Durham, NC: Duke University Press, 2010); Donna Haraway, *The Companion Species Manifesto: Dogs, People, and Significant Otherness* (Chicago: Prickly Paradigm Press, 2003); Bruno Latour, *Politics of Nature: How to Bring the Sciences into Democracy*, trans. Catherine Porter (Cambridge, MA: Harvard University Press, 2004); Quentin Meillasoux, *After Finitude: An Essay on the Necessity of Contingency*, trans. Ray Brassier (London: Continuum, 2008); Timothy Morton, *The Ecological Thought* (Cambridge, MA: Harvard University Press, 2010).

21. See Michel Foucault, "The Confession of the Flesh" (1977), interview in *Power/Knowledge: Selected Interviews and Other Writings*, ed. Colin Gordon (New York: Vintage, 1980), 194–228.

22. Latour, *Politics of Nature*, 49 and throughout; Haraway, *Companion Species Manifesto*, 1 and throughout; Lévi-Strauss, *The Savage Mind*, 246, 247.

23. Malabou, *What Should We Do with Our Brain?*, 5.

24. Chakrabarty, "The Climate of History," 222.

25. Chakrabarty, "The Climate of History," 201.

26. See, in particular, Ato Quayson, "The Sighs of History: Postcolonial Debris and the Question of (Literary) History," *New Literary History* 43, no. 2 (spring 2012) and the other essays in this special edition responding to Chakrabarty's "The Climate of History" and Robert Young's "Postcolonial Remains" (published in the previous edition of the journal, *New Literary History* 43, no. 1 [winter 2012]: 19–42). In addition to my comments on this aspect of Chakrabarty's thought in the preface and introduction to this book, I have also addressed this issue at length in my essay "The Human Shore: Postcolonial Studies in an Age of Natural Science," *History of the Present* 2, no. 1 (spring 2012): 1–23.

27. Chakrabarty, *Provincializing Europe*, 6.

28. Chakrabarty, *Provincializing Europe*, 70.

29. Chakrabarty, *Provincializing Europe*, 7.

30. Chakrabarty, *Provincializing Europe*, 15, 14.

31. Chakrabarty, *Provincializing Europe*, 16.

32. Chakrabarty, *Provincializing Europe*, 66.

33. Chakrabarty, "The Climate of History," 212.

34. Chakrabarty, "The Climate of History," 221, 216. See Robert Young, "Postcolonial Remains," *New Literary History* 43, no. 1 (winter 2012): 19–42.

35. Chakrabarty, "The Climate of History," 220.

36. Chakrabarty, "The Climate of History," 208.

37. Chakrabarty, "The Climate of History," 218.

38. Chakrabarty, "The Climate of History," 210.

39. In my first effort to consider the Anthropocene as the new situation of postcolonial studies, that is very much what I tried to do. See Ian Baucom, "The Human Shore: Postcolonial Studies in an Age of Natural Science," *History of the Present* 2, no. 1 (spring 2012): 1–23.

40. Jane Bennett, "Earthling, Now and Forever?" in *Making the Geologic Now: Responses to Material Conditions of Contemporary Life*, Elizabeth Ellsworth and Jamie Kruse, eds. (New York: Punctum Books, 2013), 245–46; Chakrabarty, "The Climate of History," 210; Bruno Latour, *We Have Never Been Modern*, trans. Catherine Porter (Cambridge, MA: Harvard University Press, 1993), 142 and following; Morton, *The Ecological Thought*, 59–97.

41. Bennett, "Earthling," 246.

42. Jean Paul Sartre, *The Family Idiot (Volume One): Gustave Flaubert, 1821–1857*, trans. Carol Cosman (Chicago: University of Chicago Press, 1981), ix–x.

43. Sartre, *The Family Idiot (Volume One)*, ix.

44. Walter Benjamin, "Theses on the Philosophy of History," in *Illuminations: Essays and Reflections* (New York: Schocken, 1969).

45. Benjamin, "Theses on the Philosophy of History."

46. David Mitchell, *Cloud Atlas: A Novel* (New York: Random House, 2004), 202.

47. See Agamben, *Homo Sacer*, throughout.

48. See Giorgio Agamben, *The Open: Man and Animal* (Stanford, CA: Stanford University Press, 2004), 12; Bill McKibben, *Eaarth: Making a Life on a Tough New Planet* (New York: Macmillan, 2010).

49. Mitchell, *Cloud Atlas*, 294.

50. Mitchell, *Cloud Atlas*, 309.

51. Mitchell, *Cloud Atlas*, 308.

52. Dipesh Chakrabarty, "The Time of History and the Times of Gods," in *The Politics of Culture in the Shadow of Capital*, ed. Lisa Lowe and David Lloyd (Durham, NC: Duke University Press, 1997), 35–60.

53. Haraway, *The Companion Species Manifesto*, 20.

54. Farhad B. Idris, "Realism," in *Encyclopedia of Literature and Politics: Censorship, Revolution, and Writing, A–Z [Three Volumes]*, ed. M. Keith Booker (Wesport, CT: Greenwood, 2005), 601.

55. George Lukács, *Studies in European Realism* (New York: Grosset and Dunlap, 1964), 145.

56. Lukács, *Studies in European Realism*, 145.

57. Lukács, *Studies in European Realism*, 145–46.

58. Dipesh Chakrabarty, "Postcolonial Studies and the Challenge of Climate Change," *New Literary History* 43, no. 1 (winter 2012): 14.

59. Chakrabarty, "Postcolonial Studies and the Challenge of Climate Change," 1–2.
60. Chakrabarty, "Postcolonial Studies and the Challenge of Climate Change," 14.
61. Chakrabarty, "Postcolonial Studies and the Challenge of Climate Change," 2.
62. Chakrabarty, "Postcolonial Studies and the Challenge of Climate Change," 14.
63. Chakrabarty, "Postcolonial Studies and the Challenge of Climate Change," 14.
64. Chakrabarty, "Postcolonial Studies and the Challenge of Climate Change," 14.
65. Chakrabarty, "The Climate of History," 212.
66. Chakrabarty, *Provincializing Europe*, 6.
67. Chakrabarty, "Postcolonial Studies and the Challenge of Climate Change," 2.
68. Haraway, *The Companion Species Manifesto*.
69. Nyani Quarmyne, "Climate Change: 'We Were Once Three Miles from the Sea,'" Nyani Quarmyne Photography, https://www.nqphotography.com/index/GoooosIKLUDsiKIA, accessed June 12, 2018.

CHAPTER 3. **The View from the Shore**

1. See Carl Schmitt, *The Nomos of the Earth in the International Law of the Jus Publicum Europaeum*, trans. G. L. Ulmen (New York: Telos Press, 2006).
2. See Baucom, *Specters of the Atlantic*, 213–41, 265–96.
3. Note: this is what Rob Nixon has so powerfully revealed through his pathbreaking work on "Slow Violence." See Rob Nixon, *Slow Violence and the Environmentalism of the Poor* (Cambridge, MA: Harvard University Press, 2013).
4. IPCC, Climate Change 2013, TS-64, accessed December 17, 2013.
5. IPCC, Climate Change 2013, TS-14, accessed December 17, 2013.
6. IPCC, Climate Change 2013, TS-64, accessed December 17, 2013.
7. IPCC, Climate Change 2013, TS-65, accessed December 17, 2013.
8. IPCC, Climate Change 2013, TS-65, accessed December 17, 2013.
9. IPCC, Climate Change 2013, TS-65, accessed December 17, 2013.
10. *4°C—Turn Down the Heat: Climate Extremes, Regional Impacts, and the Case for Resilience*, A Report for the World Bank by the Potsdam Institute for Climate Impact Research and Climate Analytics, June 2013, http://www-wds.worldbank.org/external/default/WDSContentServer/WDSP/IB/2013/06/14/000445729_20130614145941/Rendered/PDF/784240WP0Full00DoCONF0t00June19090L.pdf, xxi, xxi, xx.
11. *4°C—Turn Down the Heat*, xx.
12. *4°C—Turn Down the Heat*, xxiii.
13. *4°C—Turn Down the Heat*, 34.
14. *4°C—Turn Down the Heat*, 34, 35.
15. *4°C—Turn Down the Heat*, xxiii.
16. *4°C—Turn Down the Heat*, 15.
17. *4°C—Turn Down the Heat*, 15.
18. *4°C—Turn Down the Heat*, 1.
19. *4°C—Turn Down the Heat*, 56.
20. Toni Morrison, *Beloved* (Thorndike, ME: Thorndike Press, 1987), 43, 210.

21. See Sharpe, *In the Wake*, 105, 107, 18.

22. The full series can be viewed at Photographic Museum of Humanity, https://phmuseum.com/nyaniquarmyne/story/we-were-once-three-miles-from-the-sea-c09d2397e6, accessed January 2, 2018.

23. Benjamin, "Theses on the Philosophy of History."

24. See, the Otolith Group, "Hydra Decapita," http://otolithgroup.org/index.php?m=project&id=3#, accessed, January 2, 2018. See also Paolo Bacigalupi, *The Windup Girl* (San Francisco: Night Shade Books, 2015); Amitav Ghosh, *The Hungry Tide* (New York: HarperCollins, 2005); Omar El Akkad, *American War* (New York: Knopf, 2017); Kim Stanley Robinson *New York 2140* (New York: Orbit, 2017); Derek Walcott, *Omeros* (New York: Farrar, Straus and Giroux, 1990); Edouard Glissant, *Poetics of Relation* (Ann Arbor: University of Michigan Press, 1997); David Dabydeen, *Turner* (Leeds: Peepal Tree Press, 2002); M. Nourbese Philip, *Zong* (Middleton, CT: Wesleyan University Press, 2011); Fred D'Aguiar, *Feeding the Ghosts* (New York: Ecco Press, 1999); Toni Morrison, *Beloved* (New York: Knopf, 1987).

25. For a powerful reading of the necessity of an intersecting (intersectional) analysis of poverty, race, and climate change in the aftermaths of Hurricane Katrina and Hurricane Harvey, see Charles D. Ellison, "Race and Class Are the Biggest Issues around Hurricane Harvey and We Need to Start Talking about Them," *The Root*, August 29, 2017, https://www.theroot.com/race-and-class-are-the-biggest-issues-around-hurricane-1798536183.

26. See Jean and John L. Comaroff, *Theory from the South: Or, How Euro-America Is Evolving toward Africa* (London: Routledge, 2016); Achille Mbembe, "Theory from the Antipodes: Notes on Jean and John Comaroffs' TFS." Theorizing the Contemporary, *Cultural Anthropology* website, February 25, 2012, https://culanth.org/fieldsights/272-theory-from-the-antipodes-notes-on-jean-john-comaroffs-tfs; Sarah Nuttall, *Entanglement: Literary and Cultural Reflections on Post-Apartheid* (Johannesburg: Wits University Press, 2009); Ato Quayson, *Calibrations: Reading for the Social* (Minneapolis: University of Minnesota Press, 2003), *Oxford St., Accra: City Life, the Itineraries of Transnationalism* (Durham, NC: Duke University Press), 2014, "The Sighs of History: Postcolonial Debris and the Question of (Literary) History, *New Literary History* 43, no. 2, (2012)."

27. Dipesh Chakrabarty, "Postcolonial Studies and the Challenge of Climate Change," *New Literary History* 43, no. 1 (2012): 12, https://muse.jhu.edu/, accessed June 14, 2018.

28. Sartre, *Critique of Dialectical Reason*, 169, emphasis added.

29. Sartre, *Critique of Dialectical Reason*, 169.

30. Sartre, *Critique of Dialectical Reason*, 122–23.

31. Timothy Morton, "Hyperobjects and the End of Common Sense," The Contemporary Condition [blog], March 2010, http://contemporarycondition.blogspot.com/2010/03/hyperobjects-and-end-of-common-sense.html.

32. Morton, "Hyperobjects," 9.

33. Morton, "Hyperobjects," 7.

34. Morton, "Hyperobjects," 6, 7.

35. Morton, "Hyperobjects," 1, 81, 102.

36. Morton, "Hyperobjects," 102, 104.
37. Morton, "Hyperobjects," 83.
38. Morton, "Hyperobjects," 83; Bruno Latour, *We Have Never Been Modern*, trans. Catherine Porter (Cambridge, MA: Harvard University Press, 1993), 144.
39. Gilroy, Tanner Lectures, 67.
40. Gilroy, Tanner Lectures, 38.
41. Gilroy, Tanner Lectures, 38.
42. Gilroy, Tanner Lectures, 28.
43. Gilroy, Tanner Lectures, 38.
44. Gilroy, Tanner Lectures, 36.
45. Gilroy, Tanner Lectures, 36.
46. Gilroy, Tanner Lectures, 36, emphasis added.
47. Gilroy, Tanner Lectures, 36, emphasis added.
48. Mbembe, *Critique of Black Reason*, 162.
49. Mbembe, *Critique of Black Reason*, 182.
50. Mbembe, *Critique of Black Reason*, 166.
51. Mbembe, *Critique of Black Reason*, 168.
52. Mbembe, *Critique of Black Reason*, 163.
53. Mbembe, *Critique of Black Reason*, 169.
54. Mbembe, *Critique of Black Reason*, 180, emphasis added.
55. Mbembe, *Critique of Black Reason*, 180–81, emphasis added.
56. Ayi Kwei Armah, *The Beautyful Ones Are Not Yet Born* (London: Houghton Mifflin, 1968).
57. See Ken Saro-Wiwa, *A Month and a Day: A Detention Diary* (London: Penguin, 1995); Ken Saro-Wiwa, *Genocide in Nigeria: The Ogoni Tragedy* (Port Harcourt, Nigeria: Saros International, 1992); Rob Nixon, *Slow Violence*, 103–27.
58. United Nations, "Paris Agreement," 2015, http://unfccc.int/files/essential_background/convention/application/pdf/english_paris_agreement.pdf, accessed June 1, 2017.
59. Chakrabarty, "The Climate of History," 218.
60. Morton, "Hyperobjects," 83; Sarah Nuttall, *Entanglement: Literary and Cultural Reflections on Post-Apartheid* (Johannesburg, South Africa: Wits University Press, 2009).
61. Morton, "Hyperobjects," 81.
62. Morton, "Hyperobjects," 81–82.
63. Benjamin, "Theses on the Philosophy of History," 263.
64. Francis, *Laudato Si'*, 13, 11.
65. Francis, *Laudato Si'*, 4.
66. See Judith Butler, *Precarious Life: The Powers of Mourning and Violence* (London: Verso, 2004).
67. Francis, *Laudato Si'*, 81; Pontifical Council for Justice and Peace, *Compendium of the Social Doctrine of the Church*, 462, quoted in Francis, *Laudato Si'*, 79; see also Dirk Philipsen, *The Little Big Number: How GDP Came to Rule the World and What to Do about It* (Princeton, NJ: Princeton University Press, 2015).

CODA. **The Youngest Day**

1. Immanuel Kant, *Perpetual Peace and Other Essays on Politics, History, and Morals*, trans. Ted Humphrey (Indianapolis, IN: Hackett Publishing, Kindle Edition), 53.

2. Kant, *Perpetual Peace*, 54–55.

3. Kant, *Perpetual Peace*, 101.

4. Kant, *Perpetual Peace*, 103, 105.

5. Immanuel Kant, *The Conflict of the Faculties*, trans. Mary J. Gregor (Lincoln: University of Nebraska Press, 1992), 33, 39.

6. Dipesh Chakrabarty, "The Climate of History: Four Theses," *Critical Inquiry* 35 (winter 2009): 207.

7. Judith Butler, "Critique, Dissent, Disciplinarity," *Critical Inquiry* 35 (summer 2009): 776.

8. Butler, "Critique, Dissent, Disciplinarity," 779, emphasis added.

9. Butler, "Critique, Dissent, Disciplinarity," 780, emphasis added.

10. Kant, *The Conflict of the Faculties*, 59, 23.

11. Jean-Paul Sartre, *Search for a Method*, trans. Hazel E. Barnes (New York: Vintage, 1968), 22.

BIBLIOGRAPHY

Agamben, Giorgio. *Homo Sacer: Sovereign Power and Bare Life.* Translated by Daniel Heller-Roazen. Stanford, CA: Stanford University Press, 1998.

Agamben, Giorgio. *The Open: Man and Animal.* Stanford, CA: Stanford University Press, 2004.

Archer, David. *The Long Thaw: How Humans Are Changing the Next 100,000 Years of Earth's Climate.* Princeton, NJ: Princeton University Press, 2016.

Armah, Ayi Kwei. *The Beautyful Ones Are Not Yet Born.* London: Houghton Mifflin, 1968.

Arrighi, Giovanni. *The Long Twentieth Century: Money, Power and the Origins of Our Times.* London: Verso, 1994.

Bacigalupi, Paolo. *The Windup Girl.* San Francisco: Night Shade Books, 2009.

Baucom, Ian. "The Human Shore: Postcolonial Studies in an Age of Natural Science." *History of the Present* 2, no. 1 (spring 2012).

Baucom, Ian. *Specters of the Atlantic: Finance Capital, Slavery, and the Philosophy of History.* Durham, NC: Duke University Press, 2005.

Benjamin, Walter. "Theses on the Philosophy of History." In *Illuminations: Essays and Reflections.* New York: Schocken Books, 1969.

Bennett, Jane. "Earthling, Now and Forever?" In *Making the Geologic Now: Responses to Material Conditions of Contemporary Life,* edited by Elizabeth Ellsworth and Jamie Kruse. New York: Punctum Books, 2013.

Bennett, Jane. *Vibrant Matter: A Political Ecology of Things.* Durham, NC: Duke University Press, 2010.

Butler, Judith. "Critique, Dissent, Disciplinarity." *Critical Inquiry* 35 (summer 2009).

Butler, Judith. *Precarious Life: The Powers of Mourning and Violence.* London: Verso, 2004.

Chakrabarty, Dipesh. "Climate and Capital: On Conjoined Histories," *Critical Inquiry* 41, no. 1 (autumn 2014).

Chakrabarty, Dipesh. "The Climate of History: Four Theses." *Critical Inquiry* 35 (winter 2009).

Chakrabarty, Dipesh. "The Planet: An Emergent Humanist Category, *Critical Inquiry* 46, no. 1 (autumn 2019).

Chakrabarty, Dipesh. "Postcolonial Studies and the Challenge of Climate Change." *New Literary History* 43, no. 1 (2012).

Chakrabarty, Dipesh. *Provincializing Europe: Postcolonial Thought and Historical Difference*. Princeton, NJ: Princeton University Press, 2007.

Chakrabarty, Dipesh. "The Time of History and the Times of Gods." In *The Politics of Culture in the Shadow of Capital*, edited by Lisa Lowe and David Lloyd. Durham, NC: Duke University Press, 1997.

Chandler, James. *England in 1819: The Politics of Literary Culture and the Case of Romantic Historicism*. Chicago: University of Chicago Press, 1999.

Comaroff, Jean, and John L. Comaroff. *Theory from the South: Or, How Euro-America Is Evolving toward Africa*. London: Routledge, 2016.

Connolly, William. *Neuropolitics: Thinking, Culture, Speed*. Minneapolis: University of Minnesota Press, 2002.

Crutzen, P. J. "Geology of Mankind—The Anthropocene." *Nature* 23 (2002).

Dabydeen, David. *Turner*. Leeds: Peepal Tree Press, 2002.

D'Aguiar, Fred. *Feeding the Ghosts*. New York: Ecco Press, 1999.

Damasio, Antonio. *Looking for Spinoza: Joy, Sorrow, and the Feeling Brain*. Boston: Harcourt, 2003.

El Akkad, Omar. *American War*. New York: Alfred A. Knopf, 2017.

Ellison, Charles D. "Race and Class Are the Biggest Issues around Hurricane Harvey and We Need to Start Talking about Them." *The Root*, August 29, 2017. https://www.theroot.com/race-and-class-are-the-biggest-issues-around-hurricane-1798536183.

Engels, Friedrich. "Ludwig Feuerbach and the End of Classical German Philosophy . . . With Notes on Feuerbach by Karl Marx 1845." Berlin: Verlag von J. H. W. Dietz, 1888.

Foucault, Michel. "The Confession of the Flesh." Interview. In *Power/Knowledge: Selected Interviews and Other Writings*, edited by Colin Gordon. New York: Vintage, 1980.

Ghosh, Amitav. *The Hungry Tide*. New York: HarperCollins, 2005.

Gillis, Justin. "Heat-Trapping Gas Passes Milestone, Raising Fears." *New York Times*, May 10, 2013. http://www.nytimes.com/2013/05/11/science/earth/carbon-dioxide-level-passes-long-feared-milestone.html.

Gilroy, Paul. *The Black Atlantic: Modernity and Double Consciousness*. Cambridge, MA: Harvard University Press, 1993.

Gilroy, Paul. "Suffering and Infrahumanity," Tanner Lectures on Human Values, 2014.

Glissant, Edouard. *Poetics of Relation*. Ann Arbor: University of Michigan Press, 1997.

Guha, Ramachandra. *Environmentalism: A Global History*. London: Pearson, 1999.

Guha, Ramachandra. *The Unquiet Woods: Ecological Change and Peasant Resistance in the Himalaya*. Berkeley: University of California Press, 1990.

Haraway, Donna. *The Companion Species Manifesto: Dogs, Species, and Significant Otherness*. Chicago: Prickly Paradigm Press, 2003.

Hartman, Saidiya. *Lose Your Mother: A Journey along the Atlantic Slave Route*. New York: Farrar, Straus and Giroux, 2008.

Holsey, Bayo. *Routes of Remembrance: Refashioning the Slave Trade in Ghana*. Chicago: University of Chicago Press, 2008.

Idris, Farhad B. "Realism." In *Encyclopedia of Literature and Politics: Censorship, Revolution, and Writing, A–Z*, edited by M. Keith Booker. Westport, CT: Greenwood, 2005.

Intergovernmental Panel on Climate Change. *Climate Change 2013: The Physical Science Basis*. Contribution of Working Group I to the Fifth Assessment Report of the Intergovernmental Panel on Climate Change, ed. T. F. Stocker, D. Qin, G.-K. Plattner, M. Tignor, S. K. Allen, J. Boschung, A. Nauels, Y. Xia, V. Bex, and P. M. Midgley. Cambridge: Cambridge University Press, 2014. Accessed December 17, 2013. http://www.climatechange2013.org/images/uploads/WGIAR5_WGI-12Doc2b_FinalDraft_All.pdf.

Jameson, Fredric. *A Singular Modernity: Essay on the Ontology of the Present*. New York: Verso, 2002.

Kant, Immanuel. *The Conflict of the Faculties*. Translated by Mary J. Gregor. Lincoln: University of Nebraska Press, 1992.

Kant, Immanuel. *Perpetual Peace and Other Essays on Politics, History, and Moral Practice*. Translated by Ted Humphrey. Indianapolis, IN: Hackett Publishing, 1983.

Latour, Bruno. *Facing Gaia: Eight Lectures on the New Climatic Regime*. Translated by Catherine Porter. Cambridge, UK: Polity Press, 2017.

Latour, Bruno. *Politics of Nature: How to Bring the Sciences into Democracy*. Translated by Catherine Porter. Cambridge, MA: Harvard University Press, 2004.

Latour, Bruno. *Reassembling the Social: An Introduction to Actor-Network-Theory*. Oxford: Oxford University Press, 2009.

Latour, Bruno. *We Have Never Been Modern*. Translated by Catherine Porter. Cambridge, MA: Harvard University Press, 1993.

Lévi-Strauss, Claude. *The Savage Mind*. Chicago: University of Chicago Press, 1962.

Lukács, George. *Studies in European Realism*. New York: Grosset and Dunlap, 1964.

Malabou, Catherine. *The Ontology of the Accident: An Essay on Destructive Plasticity*, translated by Carolyn Shread. Cambridge: Polity Press, 2012.

Malabou, Catherine. *What Should We Do with Our Brain?* Translated by Sebastian Rand. New York: Fordham University Press, 2008.

Malabou, Catherine, and Adrian Johnston, eds. *Self and Emotional Life: Philosophy, Psychoanalysis, and Neuroscience*. New York: Columbia University Press, 2013.

Marx, Karl. *Later Political Writings*. Translated and edited by Terrell Carver. Cambridge: Cambridge University Press, 1996.

Mbembe, Achille. *Critique of Black Reason*. Translated by Laurent Dubois. Durham, NC: Duke University Press, 2017.

Mbembe, Achille. "Decolonizing Knowledge and the Question of the Archive." Accessed January 3, 2019. https://wiser.wits.ac.za/system/files/Achille%20Mbembe%20-%20Decolonizing%20Knowledge%20and%20the%20Question%20of%20the%20Archive.pdf.

Mbembe, Achille. "Theory from the Antipodes: Notes on Jean and John Comaroffs' TFS." Theorizing the Contemporary, *Fieldsights*, February 25, 2012. https://culanth.org/fieldsights/theory-from-the-antipodes-notes-on-jean-john-comaroffs-tfs.

McKibben, Bill. *Eaarth: Making Life on a Tough New Planet.* New York: Henry Holt, 2010.

Meillasoux, Quentin. *After Finitude: An Essay on the Necessity of Contingency.* Translated by Ray Brassier. London: Continuum, 2008.

Moore, Jason W., ed. *Anthropocene or Capitalocene: Nature, History, and the Crisis of Capitalism.* Oakland, CA: PM Press/Kairos, 2016.

Morrison, Toni. *Beloved.* New York: Alfred A. Knopf, 1987.

Morton, Timothy. *The Ecological Thought.* Cambridge, MA: Harvard University Press, 2010.

Morton, Timothy. *Hyperobjects: Philosophy and Ecology after the End of the World.* Minneapolis: University of Minnesota Press, 2013.

Morton, Timothy. "Hyperobjects and the End of Common Sense." The Contemporary Condition, March 2010. http://contemporarycondition.blogspot.com/2010/03/hyperobjects-and-end-of-common-sense.html.

Nixon, Rob. *Slow Violence and the Environmentalism of the Poor.* Cambridge, MA: Harvard University Press, 2013.

Nuttall, Sarah. *Entanglement: Literary and Cultural Reflections on Post-Apartheid.* Johannesburg: Wits University Press, 2009.

Philip, M. Nourbese. *Zong!.* Middleton, CT: Wesleyan University Press, 2011.

Philipsen, Dirk. *The Little Big Number: How GDP Came to Rule the World and What to Do about It.* Princeton, NJ: Princeton University Press, 2015.

Pope Francis. *Laudato Si': On Care for Our Common Home.* Vatican City, Italy: Libreria Editrice Vaticana, 2015.

Povinelli, Elizabeth A. *Geontologies: A Requiem to Late Liberalism.* Durham, NC: Duke University Press, 2016.

Quayson, Ato. *Calibrations: Reading for the Social.* Minneapolis: University of Minnesota Press, 2003.

Quayson, Ato. *Oxford St., Accra: City Life, the Itineraries of Transnationalism.* Durham, NC: Duke University Press, 2014.

Quayson, Ato. "The Sighs of History: Postcolonial Debris and the Question of (Literary) History." In *New Literary History* 43, no. 2 (spring 2012).

Rose, Nikolas, and Joelle M. Abi-Rached. *Neuro: The New Brain Sciences and the Management of Mind.* Princeton, NJ: Princeton University Press, 2013.

Saro-Wiwa, Ken. *Genocide in Nigeria: The Ogoni Tragedy.* Port Harcourt, Nigeria: Saros International Publishers, 1992.

Saro-Wiwa, Ken. *A Month and a Day: A Detention Diary.* London: Penguin, 1995.

Sartre, Jean Paul. *Critique of Dialectical Reason.* London: Verso, 2006.

Sartre, Jean Paul. *The Family Idiot: Gustave Flaubert, 1821–1857, Volume 1.* Translated by Carol Cosman. Chicago: University of Chicago Press, 1981.

Sartre, Jean Paul. *Search for a Method.* Translated by Hazel E. Barnes. New York: Vintage, 1968.

Schmitt, Carl. *The Nomos of the Earth in the International Law of the Jus Publicum Europaeum.* Translated by G. L. Ulmen. New York: Telos Press, 2006.

Sharpe, Christina. *In the Wake: On Blackness and Being.* Durham, NC: Duke University Press, 2016.

Sherman, David. *Sartre and Adorno: The Dialectics of Subjectivity.* New York: SUNY Press, 2008.

Smail, Daniel Lord. *On Deep History and the Brain.* Berkeley: University of California Press, 2007.

Smallwood, Stephanie E. *Saltwater Slavery: A Middle Passage from America to American Diaspora.* Cambridge, MA: Harvard University Press, 2008.

Stanley Robinson, Kim. *New York 2140.* New York: Orbit, 2017.

United Nations. "Paris Agreement." 2015. http://unfccc.int/files/essential_background/convention/application/pdf/english_paris_agreement.pdf, accessed June 1, 2017.

van Vuuren, Detlef P., Jae Edmonds, Mikiko Kainuma, Keywan Riahi, Allison Thomson, Kathy Hibbard, George C Hurtt, et al. "The Representative Concentration Pathways: An Overview." In *Climatic Change.* New York: Springer, 2011.

Walcott, Derek. *Omeros.* New York: Farrar, Straus and Giroux, 1990.

World Bank. "4° — Turn Down the Heat: Why a 4°C Warmer World Must Be Avoided." November 2012. http://climatechange.worldbank.org/sites/default/files/Turn_Down_the_heat_Why_a_4_degree_centrigrade_warmer_world_must_be_avoided.pdf.

World Bank. "4°C—Turn Down the Heat: Climate Extremes, Regional Impacts, and the Case for Resilience." June 2013. http://www-wds.worldbank.org/external/default/WDSContentServer/WDSP/IB/2013/06/14/000445729_20130614145941/Rendered/PDF/784240WP0Full00D0CONF0t00June19090L.pdf.

Young, Robert. "Postcolonial Remains." *New Literary History* 43, no. 1 (winter 2012).

Zalasiewicz, Jan, Mark Williams, Alan Haywood, and Michael Ellis. "Introduction: The Anthropocene: A New Epoch of Geological Time?" *Philosophical Transactions of the Royal Society* 369, no. 1938 (2011).

Žižek, Slavoj. *Living in the End Times.* New York: Verso, 2011.

Zunshine, Lisa, ed. *Introduction to Cognitive Cultural Studies.* Baltimore: Johns Hopkins University Press, 2010.

Zunshine, Lisa. *Why We Read Fiction: Theory of Mind and the Novel.* Columbus: Ohio State University Press, 2006.

INDEX

Adjawtor, Anikor, 82–83
African political elites, slavery and, 2
African states, climate change predictions for, 78–79
Agamben, Giorgio, 3, 21
Akkad, Omar El, 83
Akwamu people, 2
American War (Akkad), 83
Anikor, Miyorhokpor, 82–83
Anthropocene: Black Atlantic theory and, 22–26, 31–32, 86–101; Chakrabarty's critique of, 35, 52–53; climate change and, 10–12, 120n12; definitions of epoch and, 123n12; freedom and, 23–28, 33–34, 47–50, 104–9; history and, 30–32, 42–43, 68–72; materialist discourse on, 15–16, 75–79; in Mitchell's *Cloud Atlas*, 59–66; personhood in, 91–92; philosophy and, 114–17; planetary conjecture and, 6–8; postcolonial studies and, 16–18; theoretical origins of, 9–13; totality of, 72
Archer, David, 10–12
Armah, Ayi Kwei, 103–4
Arrighi, Giovanni, 2
Asante people, 2
Assessment Report 5 (AR5) (IPCC), 18, 75–76, 80–81, 87

Bacigalupi, Paolo, 83
Balzac, Honoré de, 66–67
Beasts of the Southern Wild (film), 84
The Beautyful Ones Are Not Yet Born (Armah), 103–4

Beloved (Morrison), 80, 85, 104
Benjamin, Walter, 31, 54–58, 62–63, 83, 108
Bennett, Jane, 15, 41, 47, 49–50, 65–66, 108
Berlin, Isaiah, 101
biography, Lévi-Strauss's discussion of, 38
biopolitical imperative, 108
Bizensone Fairs, 2
Black Atlantic theory: Anthropocene discourse and, 31–32, 86–101; black body in, 85–86; climate change and, 24–28, 92; freedom and humanism and, 27–28, 32–34; in postcolonial studies, 8, 13, 16–17, 92; postcolonial studies and, 74–75; slavery and, 80
Blackness, Anthropocene and, 21–28
Butler, Judith, 108, 114–16

Cesaire, Aime, 24
Chakrabarty, Dipesh: Anthropocene discourse and, 6–7, 15–16, 28, 30, 35, 52–53, 68–72; Benjamin and, 54, 64; on climate change, 7, 51, 67–68, 88–90; on force-of-the-secular, 108; on freedom, 47–50, 104; History 1 and History 2 concept of, 44–50, 59–66, 70–72, 105; planetary humanism and, 32, 87–92; on temporal order, 51
Chandler, James, 40
chronological coding, Lévi-Strauss's discussion of, 35–38
climate change: Black Atlantic and, 24–28, 92; boundary conditions for, 10–12, 120n12; Fifth Assessment Report on, 75–79; hyperobjectivity of, 94–97; postcolonial studies and, 3–8; in Sartre's *Critique*, 92–94

"Climate of History" (Chakrabarty), 44
"The Climate of History: Four Theses" (Chakrabarty), 7, 51, 67–72
Cloud Atlas (Mitchell), 57–67
Collingwood, Luke (Capt.), 2–3, 80
Comaroff, Jean, and John Comaroff, 86–87
The Companion Species Manifesto (Haraway), 66
conflict of faculties, Kant's discussion of, 114–17
correlationism, 95
cosmological assemblage: Benjamin's work and, 55–56; Mbembe on, 97–98
Critique of Black Reason (Mbembe), 21–22, 25–26, 97–98, 100–101
Critique of Dialectical Reason (Sartre), 30–32, 37, 93–94, 115–17
Crutzen, Paul, 9–10, 42

Dabydeen, David, 85
D'Aguiar, Fred, 85
"Decolonizing Knowledge and the Question of the Archive" (Mbembe), 25
deforestation, Sartre's discussion of, 93–94
Discourse on Colonialism (Cesaire), 24
dispositif, Foucault's concept of, 42–43

The Ecological Thought (Morton), 109
economic development, 108
Edoro, Ainehi, 116
The Eighteenth Brumaire of Louis Bonaparte (Marx), 17, 29–30, 37, 44
"The End of All Things" (Kant), 110–17
Enlightenment: Chakrabarty's critique of, 45–50, 68–72; in Mitchell's *Cloud Atlas*, 59–63
entanglement, in climate change discourse, 107–8
existentialism, 29

The Family Idiot (Sartre), 51–52, 63–66
Fanon, Frantz, 32, 97–100
Feeding the Ghosts (D'Aguiar), 85
Flaubert, Gustave, 64–66
force-of-the-secular imperative, 108–9
forces and forcings, climate change and, 33–34, 120n14

Fort William (Anomabo), 2–3
Foucault, Michel, 42
four degrees Celsius (4°C) threshold, 19–21
Francis (Pope), 15, 107–8
freedom: Anthropocene and, 23–28, 33–34, 47–50, 104–9; Black Atlantic and ecologies of, 23–28, 33–34; Chakrabarty's discussion of, 47–50, 104; Kant on reason and, 110–17; Materialism I and, 102–3; in Mitchell's *Cloud Atlas*, 59–63; Sartre's discussion of, 43, 64–66
The Future of Life (Wilson), 48

Ghanaian coast: Anthropocene discourse and, 31–32; climate change and, 8, 13; sea level rise along, 101–3; slavery on, 1–4
Ghosh, Amitav, 83
Gilroy, Paul, 1–2, 23–28, 32, 92, 97–99
Glissant, Edouard, 85, 87
Gold Coast slave factories, 1–3
Guha, Ramachandra, 6

Haraway, Donna, 15, 41–42, 66, 71, 98
Hartman, Saidiya, 1–2
"Heat-Trapping Gas Passes Milestone, Raising Fears" (*New York Times*), 5–6
history: Anthropocene and, 16; Benjamin's discussion of, 54–56; Black Atlantic and, 15; Chakrabarty's History 1 and History 2 discourse and, 44–50, 59–63, 70–72; Lévi-Strauss on coding of, 35–40; Marx's discussion of, 37; poststructuralism and, 29–30; race and, 25–26; Sartre's discussion of, 37
History 1 and History 2, Chakrabarty's concept of, 44–50, 59–63, 70–72, 105
History 3 concept, 46–50, 68–72
History 4°: Benjamin's work and, 55–56; Chakrabarty and concept of, 51–52; climate change and, 70–72; parliament of things and, 66–67; Sartre's work and, 51–54
"History and Dialectic" (Lévi-Strauss), 42–43, 50, 52, 56
Holsey, Bayo, 1
humanism: black critical theory and, 26–27, 98–101; Chakrabarty's discussion of, 44; ecological and cosmological perspectives in, 41–42

The Hungry Tide (Ghosh), 83
Hydra Decapita (video installation), 84, 88–89
hyperobjectivity, 32, 94–97, 101, 107, 109
Hyperobjects (Morton), 109

Idris, Farhad B., 66
Intergovernmental Panel on Climate Change (IPCC), 18–20, 75–76, 80–81, 87
International Commission on Stratigraphy, 10
International Geosphere-Biosphere Programme, 9, 42–43
International Union of Geological Sciences, 10, 12
interobjectivity, climate change and, 96, 108
In the Wake (Sharpe), 80

Jameson, Frederic, 6, 40, 93

Kant, Immanuel, 101, 110–17
Kusietey, Collins, 4, 9, 12, 28, 32, 56–57, 72–74, 81, 90–93, 103–4, 109

late capitalism, Anthropocene and, 22–24
Latour, Bruno, 15, 41–43, 52, 108
Laudato Si' (Pope Francis), 107–8
Lévi-Strauss, Claude: Anthropocene discourse and, 73–74, 108–9; Benjamin and, 56; on evolution, 41–43; Sartre's debate with, 29, 35–41, 43–44, 64–66, 116; savage dialectic of, 42–43, 50, 52, 56, 64–66
Lukács, Georg, 66–67

Malabou, Catherine, 41, 43
Marx, Karl, 8–9, 17, 29–30, 37, 44–45, 116
materialism: Anthropocene discourse and, 15–16, 32, 96–97; Chakrabarty's History 1 and 2 and, 47–50; Lévi-Strauss's historical coding and, 41
Materialism I, 25–27, 32, 47, 91, 96–99, 100–102, 105–9
Materialism II, 25–27, 32, 47, 91, 95, 97, 100–102, 105–8
Mbembe, Achille, 15, 21–28, 32, 87, 97–100
McKibben, Bill, 21, 28
Meillasoux, Quentin, 41–42
merchant agents, slavery and, 2

Mitchell, David, 57–67, 92
Moore, Jason W., 22–23
Morrison, Toni, 80, 85, 104
Morton, Timothy, 15, 32, 41, 43, 52, 65, 93–97, 107, 109

national security imperative, 108
neuroscience, history and, 41
New York 2140 (Robinson), 83–84
Nixon, Rob, 6, 17
nomological code, in Anthropocene discourse, 3–34, 55–56, 61–62, 71–72, 74, 81
now-being (*jetztzeit*) of historical time, Benjamin's concept of, 31–32
Nuttall, Sarah, 87, 107

Ogoni people, 103–4
Omeros (Walcott), 84–85
Otolith Group, 84, 88–89

Paris Climate Accord, 102
Perpetual Peace (Kant), 110–17
Philip, M. Nourbese, 85
Philipsen, Dirk, 108
Philosophical Transactions, 12
philosophy, Anthropocene discourse and, 114–17
planetary conjuncture, 23–29, 32, 87–89, 92, 101–2
Poetics of Relation (Glissant), 85
postcolonial studies: Black Atlantic in, 16–17, 74–75; Chakrabarty's discussion of, 44, 68–72; climate change and, 3–8, 80–81
"Postcolonial Studies and the Challenge of Climate Change" (Chakrabarty), 70–72
posthumanism, 98
poststructuralism, 29
Povinelli, Elizabeth, 6
power, history and scale of, 36–39
Provincializing Europe: Postcolonial Thought and Historical Difference (Chakrabarty), 40, 45–50

Quarmyne, Nyani, 3–4, 27–28, 32–33, 56–57, 73, 80–85, 87–92, 100–103, 105–6
Quayson, Ato, 17–18, 87

reason, Mbembe's critique of, 21–27
Representative Concentration Pathways (RCPS), 18–21, 75
Robinson, Kim Stanley, 83–84
Royal Dutch Shell, 103–4

Saro-Wiwa, Ken, 103–4
Sartre, Jean Paul: Anthropocene discourse and, 115–17; Benjamin and, 55; History 4° and, 51–54, 92–96; Lévi-Strauss's debate with, 29–32, 36–41, 43–44, 63–66
The Savage Mind (Lévi-Strauss), 35–40, 43, 109, 116
Schmitt, Carl, 74
Scott, Walter, 66–67
Search for a Method (Sartre), 44, 116–17
Sharpe, Christina, 13, 23, 32, 80
slavery: African political elites and, 2; Anthropocene and role of, 13–15; Ghanaian coast as staging for, 1–2; merchant agents and, 2; postcolonial theory and, 80
Smail, Daniel Lord, 41
Smallwood, Stephanie, 1
Sonmi~451 (*Cloud Atlas* character), 57–67, 71, 91–92
Southeast Asia, climate change predictions for, 76–79
spaces-of-flow, 2
Specters of the Atlantic (Baucom), 79–80, 85
Spinoza, Baruch, 49
Stoermer, Eugene F., 9–10
structuralism, 29

Taylor, Jason deCaires, 84, 86–87
temporal order, Anthropocene discourse and, 51, 79–80
Theory from the South: How EuroAmerica Is Evolving toward Africa (Comaroff and Comaroff), 86–87
"Theses on Feuerbach" (Marx), 8, 116
Tolstoy, Leo, 67
totalizing method (Sartre), 30–31
Turn Down the Heat, 76–77
Turner (Dabydeen), 85

Vicissitudes (Taylor), 84, 86–87
violence, Fanon's discussion of, 99–101

Walcott, Derek, 84–85
We Have Never Been Modern (Latour), 108
"We Were Once Three Miles from the Sea" (Quarmyne), 3–5, 27–28, 32–33, 81–85, 89–92, 100, 104–6
Wilson, Edward O., 48
The Windup Girl, 83
World Bank, 19–20, 76–77, 80–81, 87

Young, Robert, 47

Zalasiewicz, Jan, 10, 12
Zong! (Philip), 85
Zong (slave ship), 2–3, 80

www.ingramcontent.com/pod-product-compliance
Lightning Source LLC
Chambersburg PA
CBHW070359240426
43671CB00013BA/2570